Soft Matter

Edited by
G. Gompper and M. Schick

Volume 1: Polymer Melts and Mixtures

1. Polymer Dynamics in Melts
   Andreas Wischnewski and Dieter Richter

2. Self-Consistent Field Theory and Its Applications
   Mark W. Matsen

3. Comparison of Self-Consistent Field Theory and Monte Carlo Simulations
   Marcus Müller

Volume 2: Complex Colloidal Suspensions

1. Phase Behavior of Rod-Like Viruses and Virus–Sphere Mixtures
   Zvonimir Dogic and Seth Fraden

2. Field Theory of Polymer–Colloid Interactions
   Erich Eisenriegler

3. Rod-Like Brownian Particles in Shear Flow
   Jan K. G. Dhont and Wim J. Briels

Volume 3: Colloidal Order: Entropic and Surface Forces

1. Entropic Attraction and Ordering
   Randall D. Kamien

2. Phase Transitions in Two-Dimensional Colloidal Systems
   Hans-Hennig von Grünberg, Peter Keim, and Georg Maret

3. Colloids on Patterned Substrates
   Clemens Bechinger and Erwin Frey

4. Inhomogeneous Platelet and Rod Fluids
   Ludger Harnau and Siegfried Dietrich

Volume 4: Lipid Bilayers and Red Blood Cells

1. Simulations and Models of Lipid Bilayers
   Sagar A. Pandit and H. Larry Scott

2. Red Blood Cell Shapes and Shape Transformations: Newtonian Mechanics of a Composite Membrane
   Gerald Lim H. W., Michael Wortis, and Ranjan Mukhopadhyay

Soft Matter

Volume 4: Lipid Bilayers and Red Blood Cells

Edited by
Gerhard Gompper and Michael Schick

WILEY-
VCH

WILEY-VCH Verlag GmbH & Co. KGaA

Editors

Prof. Dr. Gerhard Gompper
Institute of Solid State Research
Research Centre Jülich
52425 Jülich
Germany

Prof. Dr. Michael Schick
Department of Physics
University of Washigton
P.O. Box 351560
Seattle, WA 98195-1560
USA

Cover illustration:

The membranes of almost all cells include a plasma membrane consisting of a lipid bilayer with protein inclusions. The background shows a snapshot of a typical lipid bilayer generated by a Molecular Dynamics simulation. The outer regions (green, red) typically charged or dipolar, while the interior region (blue) consists of non-polar hydrocarbon chains. The highly flexible and disordered nature of the lipid bilayer is essential for proper biological functioning of membrane protein.
The membrane of red blood cells is more complex, since it is composed of a plasma membrane plus a closely-associated elastic protein net – called the membrane skeleton – attached on the inside. Normal human red blood cells are soft bi-concave disks. Chemical and physical stresses can induce modification of this shape. Shown in the foreground is a so-called echinocytic shape. Echinocytes occur as a minority population in normal blood. (Original pictures courtesy of Sagar Pandit, See-Wing Chiu, and Gerald Lim H. W.)

All books published by Wiley-VCH are carefully produced. Nevertheless, authors, editors, and publisher do not warrant the information contained in these books, including this book, to be free of errors. Readers are advised to keep in mind that statements, data, illustrations, procedural details or other items may inadvertently be inaccurate.

Library of Congress Card No.: applied for

British Library Cataloguing-in-Publication Data
A catalogue record for this book is available from the British Library

Bibliographic information published by the Deutsche Nationalbibliothek
The Deutsche Nationalbibliothek lists this publication in the Deutsche Nationalbibliografie; detailed bibliographic data are available in the Internet at http://dnb.d-nb.de.

© 2008 WILEY-VCH Verlag GmbH & Co. KGaA, Weinheim

All rights reserved (including those of translation into other languages). No part of this book may be reproduced in any form – photoprinting, microfilm, or any other means – transmitted or translated into a machine language without written permission from the publishers. Registered names, trademarks, etc. used in this book, even when not specifically marked as such, are not to be considered unprotected by law.

Printed in the Federal Republic of Germany
Printed on acid-free paper

Cover Design SCHULZ Grafik-Design, Fußgönheim
Typesetting Da-TeX Gerd Blumenstein, Leipzig
Printing betz-druck GmbH, Darmstadt
Binding Litges & Dopf Buchbinderei GmbH, Heppenheim

ISBN 978-3-527-31502-4

Preface

This is the fourth volume in the series "Soft Matter" and the first to be devoted to biological systems, the study of which has become one of the most intense activities in soft condensed matter in recent years.

Both chapters of this volume address the properties of lipid bilayers, a system which forms the basis of all biological membranes. At first glance, however, the two contributions are very different.

The first chapter, authored by Sagar Pandit and H. Larry Scott, is concerned with the behavior of the bilayer on the scale of the individual lipid molecules. They review, therefore, the numerous microscopic models which are used to describe bilayers, and the methods by which they are simulated, in particular molecular dynamics, Monte Carlo, and Langevin dynamics. *Thermal fluctuations are important* because the hydrocarbon chains are very flexible, and accordingly their conformations are dominated by entropy. The focus is on the dynamics of the several different components which make up the bilayer. Here, the three major players are cholesterol, lipids whose chains are fully saturated, and other lipids whose chains often contain one double bond, but occasionally as many as six. *Chemical details matter*. For example, cholesterol contains a rigid multi-ring structure of which one face is "smooth" the other "rough". These interact differently with the chains. Some lipids, particularly those readily synthesized and available commercially, have two identical tails. In contrast, biological lipids often have one tail which is saturated and the other mono-unsaturated. What are the differences expected between laboratory and biological systems? How do the differences in the structures of the membrane components account, if at all, for "rafts", the putative agglomeration of cholesterol and saturated lipids which float, like rafts, in a sea of unsaturated lipids?

In contrast to the above, the chapter by Gerald Lim H. W., Michael Wortis, and Ranjan Mukhopadhyay, treats the membrane, consisting of the lipid bilayer and its associated skeleton, as a continuum surface described by various elastic modulii. *Details of the bilayer components do not matter*. They are relevant only in determining the actual values of the elastic modulii. Attention here is on the *equilibrium* shape of the red blood cell, whose characteristic size is on the order of several microns, a thousand times larger than the characteristic size of the lipid components. For the most part, *thermal fluctuations are not important* as the characteristic energy of the system is that of the bending modulus, about 50 kT. The focus of this chapter is the fascinating behavior of the shape of red blood cells, which is normally that of a flattened, biconcave disc, a "discocyte", under varying conditions. By the application of suitable chemical agents, this shape can be made to undergo several transformations: to become either more concave and invaginated (shapes denoted stomatocytes, from the Greek for "mouth"), or to exhibit external perturbations and protrusions (denoted echinocytes, from the Greek for "hedgehog"). What the authors show conclusively is that an energy functional, which accounts for curvature and stretching elasticity as well as the effect of a difference in area between the bilayer leaves, leads to the sequence of shapes normally observed, and makes many predictions about others.

There are also several striking similarities between the chapters however. First one notes the central role of Newton's laws. Of course molecular dynamics is the sequential application of Newton's laws to the components of the system. Similarly elasticity theory is the application of Newton's laws to a continuous body. What makes the application to membranes so fascinating is that, because the membrane can change shape, one must implement elasticity theory in a manner applicable to arbitrarily curved surfaces. This immediately brings us to the applications of differential geometry, which are often relegated to courses on general relativity and astrophysics. It is a pleasure to see them applied here to more terrestial problems. The necessary material is clearly presented in a masterful series of appendices. Both chapters also show that, to extract useful results, one has to rely on numerical solutions. Indeed the explication of the means to undertake this forms a large part of the chapter by Pandit and Scott.

Both of these chapters represent contemporary studies in biological systems, but they also represent complementary qualities that we hope to showcase in this series. Simulations of large biological systems are changing rapidly as the capabilities of computers increase. The chapter by Pandit and Scott presents a snapshot of the state of such simulations at this moment in time. In several years, the applications illustrated in such a chapter, and some of the underlying methodology, will probably be significantly different. The chapter by Lim, Wortis, and Mukhopadhyay on the other hand is a definitive monograph. There may be some adjustment of parameters in the future, and further comparison with experiment, but it is likely that this work will remain the definitive text. We are both pleased, and proud, to present these two outstanding contributions to the community.

March 2008　　　　　　　　　　　　　　　Gerhard Gompper and Michael Schick

Contents

List of Contributors *XI*

1 Simulations and Models of Lipid Bilayers *1*
Sagar A. Pandit and H. Larry Scott

1.1	Introduction *1*	
1.2	Atomistic Models *6*	
1.2.1	Classical Approximation *6*	
1.2.2	Molecular Dynamics *8*	
1.2.2.1	Forcefields for Lipid Simulations *9*	
1.2.2.2	Simulation Considerations and Techniques *15*	
1.2.3	Molecular Dynamics Simulations of Lipid Bilayers *19*	
1.2.3.1	System Construction and Simulation Design *20*	
1.2.3.2	Analysis and Comparison with Experiments *25*	
1.2.3.3	Surface Potential Experiments *31*	
1.2.3.4	Radial Distribution Functions *33*	
1.2.3.5	Heterogeneous Membrane Simulations *37*	
1.2.3.6	Simulations of Ordered Lipid Phases *44*	
1.2.3.7	Simulations of Asymmetric Lipid Bilayers *46*	
1.2.4	Equilibrium Monte Carlo Methods *48*	
1.2.5	Monte Carlo Studies of Lipids *50*	
1.2.6	Thermodynamic Quantities, Limitations of Atomistic Simulations *58*	
1.3	Coarse Grain Models *59*	
1.3.1	Simulations Based on Reduced "Pseudo-Molecular" Models *59*	
1.3.2	Continuum Models *60*	
1.3.3	MD Based Langevin Dynamics and Mean Field Theory *64*	
1.4	Summary *74*	
	References *76*	

Soft Matter, Vol. 4: Lipid Bilayers and Red Blood Cells
Edited by G. Gompper and M. Schick
Copyright © 2008 WILEY-VCH Verlag GmbH & Co. KGaA, Weinheim
ISBN: 978-3-527-31502-4

2	Red Blood Cell Shapes and Shape Transformations: Newtonian Mechanics of a Composite Membrane *83*

Gerald Lim H. W., Michael Wortis, and Ranjan Mukhopadhyay

2.1	Introduction *84*
2.1.1	Overview and History *84*
2.1.2	Structure of the Erythrocyte: the Composite Membrane *89*
2.1.3	What Fixes the Area and Volume of the Red Cell? Flaccid vs. Turgid Cells *90*
2.1.4	Shape Determination for a Flaccid Red Cell at Equilibrium: Membrane-Energy Minimization *93*
2.1.5	Ingredients of the Membrane Shape-Energy Functional $F[S]$ *94*
2.1.6	Shape Classes, Stability Boundaries and Phase Diagrams *95*
2.1.7	Understanding the SDE Transformation Sequence: Universality and the Bilayer–Couple Hypothesis *98*
2.1.8	Perspective and Outline *100*
2.2	Structure of the Cell Membrane; the SDE Sequence *101*
2.2.1	Plasma Membrane *102*
2.2.2	Membrane Skeleton *104*
2.2.3	More on the SDE Sequence of Cell-Shape Transformations *105*
2.3	Membrane Energetics *107*
2.3.1	Energies of Constraint *108*
2.3.2	Bending Energy of the Plasma Membrane *109*
2.3.3	Elastic Energy of the Membrane Skeleton *113*
2.3.4	Dimensionless Variables and Scaling *117*
2.3.5	History: Other Red-Cell Models *119*
2.4	Equations of Membrane Shape Mechanics *122*
2.4.1	Introduction *122*
2.4.2	Mechanics of the Plasma Membrane *124*
2.4.2.1	Fluid Membrane Without Bending Rigidity *125*
2.4.2.2	General Equilibrium Conditions for Membranes with Internal Stresses *128*
2.4.2.3	Fluid Membrane with Bending Rigidity *131*
2.4.3	Mechanics of the Membrane Skeleton *136*
2.5	Calculating Shapes Numerically *138*
2.5.1	Construction of an Initial Spherical Net $\widetilde{S}_{\mathrm{sphere}}$ *139*
2.5.2	Discretization of $F_{\mathrm{con}}[S]$ *140*
2.5.3	Discretization of $F_{\mathrm{pm}}[S]$ *141*
2.5.4	Discretization of F_{ms} *144*
2.5.5	Energy Minimization by the Metropolis Monte Carlo Algorithm *145*

2.6	Predicted Shapes and Shape Transformations of the RBC	147
2.6.1	Shape Classes	148
2.6.2	Shape Transitions, Trajectories and Hysteresis	150
2.6.3	Phase-Trajectory Diagrams	153
2.6.4	Individual Shape Classes and Stability Diagrams	161
2.7	Significant Results and Predictions	180
2.7.1	Observed SDE Shape Classes all Occur	181
2.7.2	Reference Shape S_0 of the Membrane Skeleton is an Oblate Spheroid	182
2.7.3	Predicted Hysteresis and Fluctuation Effects in RBC Shape Transformations	183
2.7.4	Strain Distribution over the Membrane Skeleton	186
2.7.5	Large Thermal Fluctuations at the AD-to-E1 Boundaries	186
2.8	Discussion and Conclusions: The Future	190
2.8.1	Validation of the Bilayer–Couple Hypothesis	191
2.8.2	Generalized Phase Diagrams and Trajectories	192
2.8.3	Sensitivity of Results to Variation of Elastic Parameters	193
2.8.3.1	Effects of Varying μ	194
2.8.3.2	Higher-Order Nonlinear Elastic Terms	194
2.8.4	Understanding the Action of Shape-Change-Inducing Agents	195
2.8.5	Experimental Quantitation of $\overline{m}_0$	196
2.8.6	Effects of Lateral Inhomogeneity of the Red-Cell Membrane	198
2.8.7	Membrane Mechanics of RBCs of Other Mammals	200
2.8.8	Summary	201
Appendix A	Material Parameters and Related Experiments	202
A.1	Geometry: Cell Area and Volume	202
A.2	Plasma Membrane Moduli	203
A.3	Membrane-Skeleton Moduli	204
A.3.1	Linear Moduli μ and K_α	204
A.3.2	Nonlinear Terms	208
Appendix B	Symmetry Sets the Form of Elastic Energies	211
B.1	Local Bending Energy	211
B.2	Local Elastic Energy of Stretch and Shear	214
Appendix C	Differential Geometry and Coordinate Transformations	216
C.1	Basic Results from Differential Geometry	217
C.2	Coordinate Transformations and Covariant Notation	219
C.3	Physical Quantities	222
C.4	Variational Approach to Membrane Mechanics	223

Appendix D	Mechanical Equations of Membrane Equilibrium	*225*
D.1	Decomposition of the Stress Tensor for Membranes	*226*
D.2	Stress Tensor for the Helfrich Model	*226*
D.3	Inclusion of Gaussian Curvature	*231*
D.4	Deformation Matrix M in Curvilinear Coordinates	*234*
D.5	Stress Tensor for the Membrane Skeleton	*235*
D.6	Shape Equations Under Conditions of Axisymmetry: Some Examples	*237*
	References	*241*

Index *249*

List of Contributors

Gerald Lim H. W.
Center for Cell Analysis
and Modeling
University of Connecticut
Health Center
263 Farmington Avenue
Farmington, CT 06032-1507
USA

Ranjan Mukhopadhyay
Department of Physics
Clark University
950 Main Street
Worcester, MA 01610
USA

Sagar A. Pandit
Department of Physics, PHY 114
University of South Florida
4202 E. Fowler Ave.
Tampa, FL 33620
USA

H. Larry Scott
Department of Biological,
Chemical, and Physical Sciences
Illinois Institute of Technology
3101 S. Dearborn
Chicago, IL 60616
USA

Michael Wortis
Department of Physics, SSC-P8429
Simon Fraser University
8888 University Drive
Burnaby, BC, V5A 1S6
Canada

1
Simulations and Models of Lipid Bilayers

Sagar A. Pandit and H. Larry Scott

Abstract

Atomistic level molecular dynamics can provide insights into the structure, dynamics and thermodynamic stability of lipid membranes and of localized raft-like regions in membranes. However the challenges in the construction and simulation of accurate models of heterogeneous membranes are great. In this chapter we outline the steps needed to carry out and analyze atomistic simulations of hydrated lipid bilayers. While molecular dynamics is a method that is simple in its conceptual content, there are many subtle challenges that must be addressed in the construction of a simulation of a lipid bilayer in water. These include simulation algorithms, forcefields, boundary conditions, equilibration and others. We will discuss all of the basic requirements for the construction and running of a molecular dynamics simulation of a lipid bilayer. We then discuss how one analyzes the data presented by a simulation, in terms of experimental results and detailed structural and dynamical predictions of the simulation. In the final part of the chapter we show how the data from a molecular dynamics simulation can be used to construct a coarse grained model for the heterogeneous bilayer that can predict the lateral organization and stability of rafts at up to millisecond timescales.

1.1
Introduction

There are many well documented fields of research in structural biology to which physicists and physical chemists regularly contribute; equally interesting are fields of biological research for which the reverse is true. The emergence of order in complex systems, from basic bio-molecular building

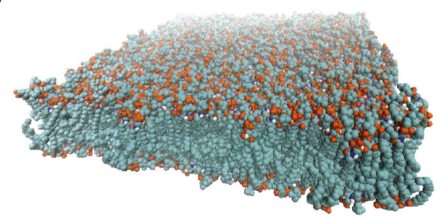

Fig. 1.1 Snapshot of a sphingomyelin lipid bilayer produced in a simulation. The lipid is 18:0 sphingomyelin, a common brain lipid. Color coding is: gray: hydrocarbon chains; red: oxygen atoms, orange: phosphorous atoms: white: hydrogen atoms and blue: nitrogen atoms. Water molecules and hydrogen atoms on hyrdocarbon chains and outermost choline groups have been excluded for clarity.

blocks (proteins, lipids, carbohydrates, sterols and others), suggests new levels of emergent material properties that hold fundamental insights in basic soft matter physics and chemistry. One such biologically inspired field in soft matter science is the study of the structure, thermodynamics, dynamics and functional behavior of biological membranes. The physics of biological membranes is uniquely interesting for multiple reasons including, but not restricted to, the following: they self-assemble spontaneously in solution, they are quasi-two-dimensional, they are composed of relatively large but flexible molecules with many intramolecular degrees of freedom, they exhibit complex phase behavior and they are capable of incorporating larger biomolecules, like proteins, without compromising their basic structural integrity. This level of structural diversity presents many challenges to those desiring to use modeling to dissect the underlying physical and chemical properties of biomembranes, and it also presents the possibility that new physics may emerge from biomembrane modeling.

Typical biomembranes are highly non-homogeneous in composition but the basic underlying structural matrix, the lipid bilayer, is the same in almost all prokaryotic and eukaryotic cells. Figure 1.1 shows a snapshot of a lipid bilayer from a simulation. The commonly accepted conceptual model for a biological membrane is such a double layer of lipid molecules, within which are embedded sterols and a bewildering variety of membrane proteins. Lipid bilayers are back-to-back monomolecular layers of phospholipid molecules. A typical phospholipid molecule, generally of a molecular weight around 750, consists of two distinct parts: a water-soluble, or hydrophilic part, and a water-insoluble, or hydrophobic part. Figure 1.2 shows a diagram of a commonly studied phospholipid, dipalmitoylphosphatidylcholine (DPPC). Figure 1.2 shows that a typical biological lipid consists of three distinct chains linked through a "backbone". In phospholipids the backbone is a three-carbon glycerol link while for sphingolipids the link is a sphingosine group. In all cases of biological interest two of the chains are made of CH_2 groups connected by single or, in some cases, double bonds and terminated by a methyl CH_3. These two chains are highly hydrophobic, with very low solubility in water. The third chain contains phosphate (PO_4) and choline ($N(CH_3)_3$) fragments connected by a methylene (CH_2). At neutral pH the phosphate carries a net negative charge that is balanced by a net positive charge on the choline. The dipolar nature of the polar chain (referred to as the "head group") renders the head group highly soluble in water. As a consequence of the amphiphilic (half water-hating and half water-loving) nature of the molecule, when dispersed in excess water lipids like DPPC self assemble into structures that shield the hydrophobic regions from water and maximize contact between the polar regions and water. One class of structures are lipid bilayers. In a lipid bilayer in an aqueous solution, the hydrophobic parts make up the interior while the hydrophilic parts make up the interface with the water. Lipid bilayers can spontaneously form closed spherical structures, or vesicles, when mixed in excess water. Vesicles are the natural "compart-

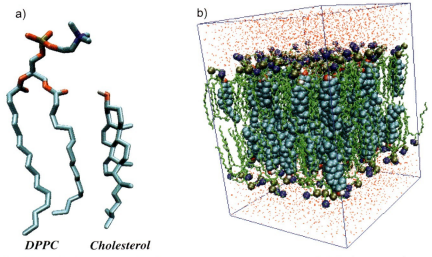

Fig. 1.2 (a) Stick models of DPPC and cholesterol molecules. In DPPC, the upper chain is the hydrophilic polar head group, and the two lower chains are the hydrophobic part of the molecule. (b) A snapshot of a simulated bilayer made up of DPPC and cholesterol. This snapshot also shows water molecules above and below the bilayer.

ments" that separate the interior from the exterior of a cell. Cholesterol, also shown in Fig. 1.2, is another biologically important lipid that contains hydrophobic and hydrophilic parts. In the case of cholesterol the hydrophobic part has four fused rings and a short tail, while the hydrophilic part consists of a single hydroxyl attached to the first ring. Of particular interest, as we will discuss later in this chapter, are the three methyls that protrude from one side of the ring portion of cholesterol. They are represented by three horizontal sticks in Fig. 1.2.

The conceptual view of biomembranes based on the lipid bilayer concept, called the fluid mosaic model, was first proposed in 1972 (Singer and Nicholson 1972). In the fluid mosaic picture, membrane proteins, sterols (such as cholesterol) and other biologically essential molecules reside in, on or penetrate the lipid bilayer, performing essential biochemical functions required by the cell. Since 1972 much progress has been made in understanding the properties of a fluid mosaic membrane at a molecular level. It has become clear that the fluid mosaic picture describes a highly dynamic structure of extremely hetreogeneous composition that can fluctuate in its lateral organization and in the ordering of the lipid chains in response to stimuli from the interior and the exterior of the cell. Within the plane of the lipid bilayer there is rapid lateral diffusion and dynamical fluctuations in structure on a sub-nanometer scale. Since a typical membrane is made up of perhaps a dozen different lipids, and contains sterol and proteins, to dissect the underlying physical interactions is a formidable problem. To better understand the many complex physical and chemical reactions and interactions which drive biological functions, a tractable approach is to first gain insight into the molecular interactions within simple lipid bilayers, such as shown in Fig. 1.1. In model membranes the composition is greatly simplified compared to that of biological membranes. Typically model membranes contain only one or two different lipids, a controlled amount of cholesterol and/or one or no membrane proteins. Figure 1.2 shows the structure of two commonly studied lipid molecules. Figure 1.2 also shows a snapshot that illustrates a typical distribution of lipids and water in a small part of a model bilayer which was generated by a simulation. The highly disordered, fluid nature of the bilayer can be seen in this figure.

Model membranes have been quantitatively studied experimentally by a wide variety of methods (Merz and Roux 1996; Nagle and Tristram-Nagle 2000; Tristram-Nagle and Nagle 2004). Over the years an interplay has developed between experiment and simulation, wherein experimental data revise and improve the quality of simulations and in turn simulations are used to interpret experimental data. As a consequence, there is now a sizable and growing data base of structural and dynamical data from which it is possible to construct theoretical models for lipid bilayers. The goal of theoretical models is to understand how the microscopic intermolecular interactions in lipid bilayers lead to the experimentally observed structures. As an insight is gained into simple lipid bilayers through this process, the goal of future modeling work is to apply the new information to the expanded study of lipid bilayers of a more complex and biological composition. Unfortunately the complex structure of even the simplest lipid molecules (see Figs. 1.1 and 1.2) makes modeling especially difficult. Each of the three chains (two in the hydrophobic region and one hydrophilic chain) can change shape by rotations about atomic bonds (dihedral rotations), so that the conformation space of a single molecule is huge. A lipid bilayer is not just a simple two-dimensional fluid but a two-dimensional fluid of molecules, each of which has a large number of internal degrees of freedom. Hence, the interaction between pairs of lipid molecules depends not only on their separation, but also in some complex fashion on the conformational shapes of the molecules.

Theoretical models for lipid bilayers that concentrated on the main chain melting phase transition have been proposed (Nagle 1973; Scott 1975; Nagle and Scott 1978). However, it is generally very difficult to realistically use the analytical tools of statistical mechanics to model a complex system such as a lipid bilayer without major approximations. The best approximation schemes are those that are guided by experimental data. This difficulty was a severe limitation in the early development of this field due to a lack of detailed understanding of atomic level interactions between lipid molecules in bilayers, an essential requirement for the construction of realistic models. However over the past 10–15 years this situation has begun to change as atomistic simulations of lipid bilayers have progressed. A major motivation for doing simulations for lipid bilayers is that by doing reliable simulations (where reliability means the simulations are of sufficient scale in size and simulation time, and that they agree with all available experimental data) one gains atomic level structural and dynamical coordinates of all atoms in the system. The basic prediction of any simulation is a trajectory or a set of system configurations consisting of atomic coordinates, velocities (for molecular dynamics simulations) and interaction potentials which can be directly linked to the macroscopic behavior observed in experiments. This wealth of atomic resolution, structural and dynamical data is then available to test hypotheses used by experimentalists in interpreting measurements and for the design of new experiments. It can also be used by theorists to formulate better coarse grained statistical mechanical models for membrane phase behavior.

In this chapter we discuss the current state of atomistic simulations of model membranes. We will also describe some ways by which information from atomistic simulations can be directly employed for improved statistical mechanics modeling of model membranes at scales that greatly exceed those of simulations. In the first section we review the basic simulation techniques, molecular dynamics (MD) and Monte Carlo (MC) as applied to model membranes, and we discuss the interplay between experiments and simulations. In Section 1.3 we discuss efforts to build upon atomistic simulations by constructing coarser grained models that can extend the scope of atomistic simulations. In Section 1.3.3 we describe models based on Mean Field Theory that predict model membrane structure at a thermodynamic scale and over milliseconds in time. Section 1.4 contains concluding discussions.

1.2
Atomistic Models

1.2.1
Classical Approximation

Atomistic simulations are simulations in which, in principle, all individual atoms are included. In practice, for systems as large as lipid bilayers it is necessary to somewhat relax this goal in several ways. Firstly, atoms (or groups of atoms) are approximated as classical particles centered at the respective atomic nuclei. Secondly, atomistic simulations of complex molecular systems like lipid bilayers are generally run under a set of key assumptions concerning the treatment of the intermolecular interactions between atoms that are chemically bonded, and atoms that are not chemically bonded.

The first assumption takes advantage of the fact that the dynamical time scales for electronic degrees of freedom are several orders of magnitude faster than the dynamical time scales for the positions and momenta of individual atomic nuclei. Under this approximation, one can include the effects of electronic motions into averaged interactions that are embedded into simpler bonded and non-bonded interactions. A complete derivation of the classical interactions in atomistic simulations within the framework of the time dependent self consistent field approximation has been presented by Marx and Hutter (2000). Of course within this approximation it is not possible to simulate chemical reactions. Also within this framework electrostatic properties of single atoms are compromised due to the lack of electronic polarizability. However, these are not severe compromises if one is only interested in physical properties as in lipid bilayer simulations. The second assumption is that all the non-bonded interactions are represented by sums of forces between pairs of atoms only, without the inclusion of triplet or higher many-body interactions.

1.2 Atomistic Models

$$V_{\text{tot}} = \sum_{\text{bonds}} K_b(r-r_0)^2 + \sum_{\text{angles}} K_\theta(\theta-\theta_0)^2$$

$$+ \sum_{\text{impropers}} K_\Phi(\Phi-\Phi_0)^2 + \sum_{\text{dihedrals}} K_\phi[1+\cos(n\phi-\phi_0)]$$

$$+ \sum_{\text{non-bonded pairs}} \left\{ \frac{q_i q_j}{r_{ij}} + \left[\frac{C_{ij}^{(12)}}{r_{ij}^{12}} - \frac{C_{ij}^{(6)}}{r_{ij}^6} \right] \right\} \quad (1.1)$$

The most common form of the intermolecular potential energy function used in simulations of lipid bilayers is shown pictorially in Fig. 1.3 and explicitly in Eq. (1.1). The sum runs over (in the order of terms in Eq. (1.1)) bonds, bond angles, improper and proper dihedrals, and all pairs of atoms that are on different molecules or are separated by more than four interatomic bonds on a molecule. The dihedral function in Eq. (1.1) represents the energy of a connected set of four consecutive atoms on a molecule. Figure 1.3 pictorially summarizes the contributions to Eq. (1.1).

Equation (1.1) shows a typical potential energy function that incorporates these assumptions, and is used in most of the classical atomistic simulations. The expressions in Eq. (1.1) represent strong approximations regarding the nature of the interatomic interactions between lipids. In reality interactions between polyatomic molecules are not spherically symmetrical, and electric

$$V_{bond} = \frac{1}{2} k_{ij}(r_{ij} - b_{ij})^2$$

$$V_{angle} = \frac{1}{2} k_{ijk}^\theta (\theta_{ijk} - \theta_{ijk}^0)^2$$

$$V_{dihedral} = \sum_n k_{ijkl}^\phi (1 + \cos(n\phi_{ijkl} - \phi_{ijkl}^{n_0}))$$

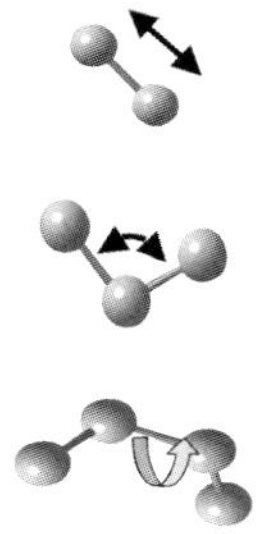

$$V_{LJ} = \sum_{i<j} \frac{C_{ij}^{(12)}}{r_{ij}^{12}} - \frac{C_{ij}^{(6)}}{r_{ij}^6} \qquad V_{Electrostatics} = \sum_{i<j} f \frac{q_i q_j}{\varepsilon_r r_{ij}}$$

$$V = V_{bond} + V_{angle} + V_{dihedral} + V_{LJ} + V_{Electrostatics}$$

$$F = -\nabla V$$

Fig. 1.3 Figure showing typical interaction functions used in atomistic molecular dynamics simulations. Bond length, angle and dihedral contributions are also shown pictorially on the right.

polarization plays a critical role, so that simple coulombic and 6–12 potential functions are not correct. However, they are necessary if one is to run a simulation of a complex system over time and length scales that compare with an experiment. The use of simplified interaction potentials can be justified by carefully tuning the parameters in Eq. (1.1) to fit independent experimental data, a process we will describe in a subsequent section.

Once a classical interaction function such as Eq. (1.1) is accepted as derived approximately from full quantum molecular systems and parameterized in accordance with multiple sets of experimental data, several sampling techniques, such as molecular dynamics, Monte Carlo, or Brownian dynamics, can be employed to generate an ensemble of statistical samples in the accessible phase space of the system. These points are then used to determine the average properties of the simulated systems. For the lipid bilayer systems the most popular simulation tools are molecular dynamics and Monte Carlo. We will examine these methodologies in the following sections.

1.2.2
Molecular Dynamics

In molecular dynamics (MD), configurations of the system are sequentially generated beginning with an initial set of positions and velocities of all the atoms in the system and integrating Newton's equations of motion using the potential function such as that in Eq. (1.1) for all the atoms. The outcome of an MD simulation is a "pseudo trajectory" of positions and velocities of all of the atoms in the simulation system, and this data set can be used to determine structural and dynamical properties of the system at a level of atomic resolution that exceeds what can be determined from other modeling methods.

In a typical molecular dynamics simulation atoms or groups of atoms are treated as point particles for the calculation of positions, momenta and intermolecular forces. Simulations begin with predetermined configurations and, usually, random velocities for all of the individual atoms. Then Newton's equations of motion are numerically integrated, where the force is obtained from the negative gradient of a potential function such as Eq. (1.1). Iterative application of this generates sequential sample points which trace classical trajectories in the phase space of the particles. Ideally molecular dynamics simulations provide a detailed description of the dynamic evolution of complex molecular systems. However, the coupled differential equations obtained from Newton's equations are a highly nonlinear system of equations. Many such systems of nonlinear differential equations are known to exhibit chaotic behavior. That is, they are deterministic systems but have long-term behavior that is practically impossible to predict due to an exponential sensitivity to the initial conditions. One of the peculiar properties of chaotic dynamics is that two almost identical starting configurations evolve very differently in time. For this reason the main use of molecular dynamics is to generate sample points of the system rather than to predict a precise final microscopic state that is found after evolution from a fixed initial state. This has strong implications for the applicability of molecular dynamics simulations. Due to the chaotic nature of molecular dynamics trajectories it is difficult to ascertain if a particular final state is representative of an experimental version of the system, or is a result that contains peculiarities that have evolved from the particular initial state. However, if one is concerned only with the overall distribution of final states rather than a particular state the chaotic nature of the simulation is an advantage, in that successive microstates diverge exponentially from each other, and thereby a sampling of microstates is gained in a long simulation. Systems that are more solid-like and evolve more slowly to sample more limited regions of configuration space are not as well simulated using molecular dynamics. Precisely for these reasons molecular dynamics is an excellent tool in the study of lipid bilayers. Lipid bilayers are in a fluid phase in biological systems, and also in most experimental model membrane studies. Then, the quantities of interest are not usually the particular conformations of lipids but instead are found as averages over many configurations of properties like densities and order parameters.

It is clear that the application of molecular dynamics to lipid bilayers is more complex than an application to a simple fluid. The simulation cell is non-uniform, with lipids in a bilayer surrounded by water. There are very strong local electrostatic forces that if not screened or otherwise balanced, will lead to instabilities. The interior of the bilayer contains hydrocarbon chains that are fairly tightly packed and strongly resist penetration by water. The evolution of a simulation under molecular dynamics strongly depends on the initial state. For all of these reasons it is essential that the intermolecular forces in the simulation accurately depict the actual interatomic forces. This is a challenge as the actual interatomic forces are extremely complex and nonlinear and so approximate force expressions, as discussed above, are implemented. In the following section we describe the process by which the interatomic force functions are defined and optimized for lipids and water in hydrated bilayer environments.

1.2.2.1 Forcefields for Lipid Simulations

The exact form of the interaction function (see Eq. (1.1)) and the set of all the parameters defined therein, are referred to as the forcefield for a molecular dynamics simulations. Equation (1.1) represents one particular forcefield and is the functional form that is almost universally used, but it is by no means unique. Due to the extensive simplifications in forcefields compared to actual interatomic interactions, simulation forcefields must be parameterized to be sufficiently complex and robust to reproduce most experimentally measured structural properties in the domain of applicability of the molecular dynamics simulation. The forcefield parameters can be grouped by type:

- Parameters for forces between bonded atoms.
- Parameters for intramolecular forces between non bonded atoms within the same molecule.
- Parameters for intermolecular forces between non bonded atoms on different molecules.

As described above, Eq. (1.1) contains terms for each type of interaction. We now turn to the determination of the parameters for each of these classes of interatomic interactions. The discussion will be primarily based on Eq. (1.1) but one should keep in mind that, depending on required accuracy and availability of resources, several additional interaction terms can be introduced. For example anharmonic bonds explicit hydrogen bond interactions could be added; for intramolecular forces special 1–4 interaction parameters (between atoms on one molecule separated by three bonds, to be discussed below) could be added; and for non-bonded interactions the more to complex Morse potential is an alternative to 6–12 Lennard–Jones potential (Cramer 2006).

Of particular concern are the 1–4 interactions between atoms on a chain within a single molecule that are separated by two other atoms on the chain. They are important in the generation of molecular conformations, as in many chain rotational states the 1 and 4 atoms in a lipid chain interact quite strongly in a bilayer environment. The simplest approach is to use dihedral interaction functions as will be described below. These potentials are designed to describe the torsion angle interactions within a chain, and this includes the 1–4 interactions. Also of concern is the set of forcefield parameters that are used for water. Clearly it is necessary to accurately describe interactions between waters and between waters and lipids, but accurate models turn out to be prohibitively expensive in terms of computational time. The pros and cons of various water models will be discussed below. Electrostatic interactions present a challenging problem to simulations because of the long range nature of the Coulomb force. This issue will also be addressed in a subsequent subsection. There is also a concern that actual interactions between complex molecules may require many-body contributions. However experience seems to indicate that the limitation of the potential function to two-body interactions is, for the simulation of lipid bilayers, a reasonable assumption provided the parameters are chosen carefully and the interactions are not truncated too rapidly.

The forcefields in use today have been constructed to address these challenges, and now simulations of simple lipid bilayers have achieved a high level of success in reproducing structural experimental properties including density, chain order and the distribution of the various atoms within the bilayer. As we will discuss in subsequent sections, the challenge is now to extend simulations to more complex and biologically relevant membranes. As an example in the following subsection we will describe the determination of a forcefield for a united atom based MD simulation of lipids.

In principle one should determine the forcefield parameters for all the interactions in Eq. (1.1) by a series of independent calculations and simulations that determine the values that provide the best fit to independent experimental data. Several research groups have calculated united atom forcefields for lipids and cholesterol (Chiu et al. 1999b; Berger et al. 1997). Other groups have focussed on all-atom forcefields for lipid simulations (MacKerell Jr 2004). However forcefields are in a continual state of updating, as new data become available, and, if a simulation of a new lipid or a cholesterol analog such as ergosterol is planned, then the following forcefield parameters from Eq. (1.1) should be calculated for the new molecule or molecules.

- Bond length and bond angle parameters.
- Effective atomic charges.
- Dihedral torsion parameters.
- Lennard–Jones, or 6–12, parameters.

The standard strategy in the general determination of the various forcefield parameters is to consider "model compounds", or "fragments", that is smaller molecules that are part of the lipid molecule and for which experimental data are available. The forcefields for the model compounds are adjusted to optimally fit available experimental data and then are incorporated into the relevant part of the lipid forcefield. To fill in the forcefield for a simulation it is necessary to specify forcefield parameters for harmonic bonds, for atomic charges, for 6–12 parameters and for torsional interactions. This must be done for each atom type in the molecule and, for inter-molecular interactions, for all pairs of atom types. For the case of chemical bonds, the complicated quantum mechanical interactions that determine bond length and angle fluctuations are modeled in Eq. (1.1) by simple harmonic potentials. In lipid bilayers around room temperature the lowest vibrational mode is generally the dominant one. For harmonic bond length and bond angle interactions parameters are adjusted to reproduce experimentally measured molecular spectra from model compounds, and this part of the forcefield is therefore quite accurate (Cramer 2006).

For atomic charges the choices are somewhat more *ad hoc*. To compute partial charges several steps are required. First density functional theory (DFT) calculations are performed on small fragments of the molecules. Then, based on the electron probability densities from DFT, partial charges are computed by performing a population analysis of the electrons. An alternative method that is popularly used to obtain partial charges is based on empirical parameters like electronegativity and ionization potential of the atoms in the molecule. This method, although less accurate, is extremely fast compared to the density functional theory method. Hence, it can be used on each sample point to generate charge distributions based on configurational geometries.

The set of torsion angle potential parameters in a linear chain modulate the interactions between quadruplets of bonded atoms along a chain, and thereby control the intramolecular configurations that are allowed. The torsional parameters required for MD can be determined using the procedure outlined by Reiling et al. (1996). In general, in this method a potential energy profile, Eq. (1.2), as a function of the dihedral angle is computed at the *ab-initio* level using a quantum chemistry package such as GAUSSIAN03 (Frisch et al. 2004) with *ab-initio* calculations at B3LYP/6-31G(d,p) level with all other geometrical parameters optimized. The general form for the torsion potential is

$$V_{\text{Tor}}(\phi) = K_\phi \left(1 + \cos(\delta)\cos(m\phi)\right) \qquad (1.2)$$

where K_ϕ, δ, m, and ϕ are the force constant, phase shift, multiplicity, and dihedral angle, respectively. The procedure based on Eq. (1.2) is employed for dihedrals in or near the polar part of the lipid molecules. For lipid hydrocarbon chains the united atom model requires a dihedral potential with multiplicity up to $m = 5$. For the hydrocarbon chain torsions, one generally uses one of the two well known and well tested dihedral potentials: Ryckaert–Bellemans (1978) (see Eq. (1.3)) or Kuwajima (1994) (see Eq. (1.4)) potential functions are used. Both of these torsional functions are computed with explicit 1–4 non-bonded interactions included. In general, the torsional parameters can be derived with or without 1–4 interactions included.

$$V_{\text{Tor}}(\phi) = 9.28 + 12.16\cos\phi - 13.12(\cos\phi)^2 - 3.06(\cos\phi)^3$$
$$+ 26.24(\cos\phi)^4 - 31.5(\cos\phi)^5 \,\text{kJ/m} \qquad (1.3)$$

$$V_{\text{Tor}}(\phi) = 7.35 + 19.40\cos\phi + 4.35(\cos\phi)^2 - 31.10(\cos\phi)^3 \,\text{kJ/m} \qquad (1.4)$$

Chiu et al. determined head group torsion parameters by a similar procedure, using a different set of model compounds to calculate parameters for separate fragments that make up the polar group. The procedure was applied to the phosphocholine head group, which consists of choline and phosphate fragments as well as two ester groups which link the head group and the hydrocarbon tails (Fig. 1.4). The following model systems were used for the determination of the dihedral parameters for these molecular sub-groups:

- Choline group: ethyltrimethylammonium;
- Phosphate group: dimethyl phosphate;
- Ester group region: methyl acetate CH_3COOCH_3.

Head group paramaterization was done using *ab initio* calculations to determine partial charges, and as MD simulations to determine non-bonded interactions for each of these compounds. The end result was a set of force-field parameters for the polar head group of DPPC and other lipids with a phosphocholine polar group.

The 6–12 parameters for hydrocarbon chain united atoms can be determined by adjusting non-bonded forcefield parameters in MD simulations to obtain correct density and heat of vaporization for linear alkanes and alkenes. Berger et al. (1997) used this approach to develop a set of 6–12 parameters based on fitting volumetric and thermodynamic data for liquid pentadecane, as an analogue for the 16-carbon chains of the dipalmitoylphosphatidylcholine (DPPC) phospholipid molecule and showed that this modification agreed very well with experimental data on the DPPC bilayers. However, fitting hydrocarbon parameters to a single length hydrocarbon chain leaves the parameters underdetermined. This is because there are multiple combinations of methyl and methylene specific volumes that will result in the correct specific volume for a linear hydrocarbon of any specific length, such as pentadecane. Chiu and co-workers (1999b, 2003) then developed hydrocarbon forcefields that provided an excellent fit for volume and heat of vaporization over a wide range of hydrocarbon lengths. This effort consisted of running MD simulations of hexane, decane, pentadecane and 5-decene. The parameters that were adjusted included non-bonded interactions between CH_1, CH_2 and CH_3 molecules. The CH_1, CH_2 and CH_3 were modeled as united atoms (no explicit hydrogens). After tuning the parameters they produced one set for each type of united atom that would, in simulations of the liquid alkanes, reproduce the liquid density and the heat of vaporization.

The final step in the development of a set of forcefield parameters is of course validation against experiment. Pandit et al. (2007b) ran extended simulations of DPPC, dioleyol phosphatidylcholine (DOPC) and palmitoyloleyol phosphatidylcholine (POPC) for this purpose. Figure 1.8 shows X-ray form factors (to be discussed in more detail later in this chapter) calculated from simulation, and compared with experimental data from the group of Nagle et al. (2006, 1997). Agreement between experiment and simulation is very good out to the third lobes in the experimental data.

14 | 1 Simulations and Models of Lipid Bilayers

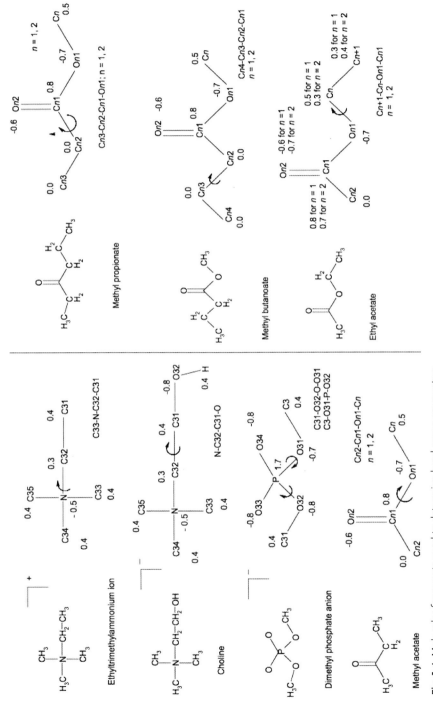

Fig. 1.4 Molecular fragments used to determine head group parameters

Table 1.1 Structural properties for DPPC, POPC, and DOPC bilayers. ^{a} V_c is calculated as $A_l \times D_c$ from reported values and $V_{HG} = V_l - V_c$. Volumes are in Å^3 and areas are in Å^2. Experimental data are from Kucerka et al. (2005a)

	DPPC simulation	DPPC expt	POPC simulation	POPC expt	DOPC simulation	DOPC expt
Temperature (K)	323	323	303	303	303	303
V_l	1212.8	1228.5	1241.8	1256	1269.7	1303
V_c	869.0	895.6^a	911.4	924.2	949.3	971
V_{HG}	343.8	332.9^a	330.4	331	320.4	331
V_{CH_2}	27.1	—	27.1	27.6	27.0	27.6
V_{CH_3}	54.2	—	54.2	53.6	54.0	53.6
$V_{CH=CH}$	—	—	43.4	44.2	43.2	44.2
A_l	64.3	64.2	66.5	68.3	68.8	72.5

Other structural data such as order parameter profiles (to be discussed later in this chapter) summarized in Table 1.1, are also in very good agreement with experimental data. The overall agreement between simulation and experiment is of sufficiently high quality that the forcefield can be used with confidence for these lipids in new simulations

1.2.2.2 Simulation Considerations and Techniques

Figure 1.5 is a flow chart for an MD simulation. As the figure shows any molecular dynamics simulation has four important components, three of which repeat many times: (i) initialization, (ii) computation of forces, (iii) numerical integration and (iv) comparison with real experimental quantities. Practical implementation of these components require use of specialized techniques, additional approximations and algorithms. Below, we discuss these components in the context of lipid bilayer simulations.

Initialization: Initialization consists of building the bilayer, adding waters of hydration and equilibrating. Each of these components will be discussed in a subsequent section. This step must be done with great care, because the trajectory of the simulation will depend on the initial state. Any unrealistic or unphysical elements present in the initial state can compromise calculated properties of the simulated bilayer.

Force computations: The intermolecular force computation step requires a significant fraction of the total CPU time in each molecular dynamics time step. For this reason numerous algorithms and strategies have been developed to increase the efficiency of intermolecular force computations. From Eq. (1.1) there are two types of forces one needs to compute: (i) bonded forces, including harmonic bond stretching forces, bond angle forces and torsional forces; (ii) non-bonded forces, including Coulombic and 6–12 forces.

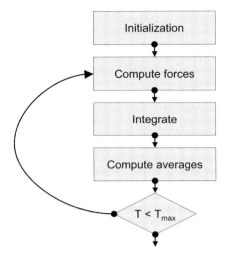

Fig. 1.5 Flow chart for molecular dynamics algorithm

In classical molecular dynamics simulations there are no chemical reactions so the bond structure of all the atoms remains invariant throughout the simulation. This fact is used in accelerating the computation of bonded interactions. Static bond lists and exclusion lists can be constructed at the begining of the simulation, and can then be used throughout simulations to compute bond length, bond angle, torsional and excluded 1–4 interaction forces. Non-bonded interactions such as Lennard–Jones and electrostatics interactions cannot be implemented as static lists because one needs to consider all possible pairs of atoms in these interactions. With periodic boundaries, these sums have to incorporate the effect of atoms from periodic images. However, if the interaction decreases with interatomic distance as fast as or faster than r^{-3} then one can use a truncation method in which the potential is calculated only up to a cutoff radius from the central atom and is considered to be zero beyond that distance. The truncation method may be simple and abrupt,

$$V(r) = \begin{cases} V_{6-12}(r) & r \leq_c \\ 0 & r > r_c \end{cases} \quad (1.5)$$

or shifted

$$V(r) = \begin{cases} V_{6-12}(r) - V_{6-12}(r_c) & r \leq r_c \\ 0 & r > r_c \end{cases} \quad (1.6)$$

Simulation software usually allows the user to select the cutoff method and the value of r_c. A conservative choice is 1.8 nm cutoff which introduces $\sim 3 \times 10^{-4}$ kJ/mol error in a typical CH_3–CH_3 interaction at cutoff distance (although many researchers use a much shorter cutoff of 1.0 nm to improve simulation efficiency but that gives rise to an error of $\sim 10^{-2}$ kJ/mol in a typical CH_3–CH_3 interaction). Neighbor lists can be constructed that store lists of atoms that are within the cutoff distance, thereby avoiding scans over all pairs of particles at each force calculation step. Because the timesteps are quite small, and lipid diffusion is also small, the neighbor list need to be updated only after a preset number of time steps, usually around 5–10. Long range electrostatic forces present a serious challenge: one can simply truncate the electrostatic interaction, but this has been shown to produce artifacts in correlation functions at the cutoff point (Patra et al. 2003), and in the molecular area per molecule in simulations since the Coulomb interaction only decreases as r^{-1}. The situation becomes even more challenging with periodic boundaries as the electrostatic interaction sum becomes conditionally convergent and requires special care in summing. A preferred choice for performing electrostatics sums in this situation is the Ewald summation algorithm. The Ewald method is described in many textbooks on computation including, for example (Cramer 2006). In software implementations the Ewald sum algorithm is further improved using Smooth Particle Mesh technique (Essmann et al. 1995). The Ewald sum method divides the electrostatic sums into real space and fourier space components. Therefore the implementation of an Ewald algorithm requires parameters such as real space cutoff, number of fourier space vectors and so on that must be chosen by the user. Most simulation programs parse the atoms in a molecule into "charge groups" for the calculation of electrostatic interactions. The user must define the charge groups in input files for a simulation. These charge groups must be neutral to avoid unwanted charge-charge correlations especially at the edges of the simulation box. Also, if the charge groups are too large, artifacts may occur in the simulation. A disadvantage of Ewald summation is that it imposes an artificial periodicity on the system. However for sufficiently large simulations this limitation does not pose a serious concern. In a study of the comparative effects of cutoffs and Ewald summations on the properties of a simulation Wohlert and Edholm (2004) found that for sufficiently large lipid simulations (over 1000 lipids plus waters of hydration) using a cutoff electrostatic interaction of 1.8 nm or greater did not affect the results of the simulation. However, for smaller simulation boxes, artifacts due to cutting off the long range Coulomb forces were apparent.

Integrators and ensembles: The propagation by molecular dynamics integration of Newtonian equations over time samples the states on a constant energy hyper–surface in the phase space of the system. The simplest algorithm for the integration step is the well known Verlet algorithm (Frenkel and Smit 2002)

$$\boldsymbol{r}(t+\Delta t) = 2\boldsymbol{r}(t) - \boldsymbol{r}(t-\Delta t) + \frac{\boldsymbol{F}(t)}{m}\Delta t^2 \qquad (1.7)$$

which has an error of order Δt^4. There are several variations of this scheme mostly used to improve performance or accuracy. The popular molecular dynamics software GROMACS (Berendsen et al. 1995) uses a method called "leap frog" Verlet to integrate trajectories. In this method velocities are evaluated at the half time step as

$$\boldsymbol{v}\left(t+\frac{\Delta t}{2}\right) = \boldsymbol{v}\left(t-\frac{\Delta t}{2}\right) + \frac{\Delta t}{m}\boldsymbol{F}(t) \qquad (1.8)$$

$$\boldsymbol{r}(t+\Delta t) = \boldsymbol{r}(t) + \boldsymbol{v}\left(t+\frac{\Delta t}{2}\right) \qquad (1.9)$$

The other variation typically used is the velocity Verlet method where each velocity and each force is computed only at each full time step. If the forces are not velocity dependent then this method requires one fewer velocity computation per time step compared to other Verlet type methods.

There are a number of higher order intergration schemes that are more accurate than the simple Verlet algorithm, such as an nth order predictor-corrector algorithm (Frenkel and Smit 2002). In this algorithm the position of a particle first n derivatives at time t are used to predict the position and its first n time derivatives at time $t + \Delta t$. Then the prediction of the second derivative is "corrected" by comparison with a calculation of the force, and the correction is used to update the other predicted derivatives at the new time. This procedure has the advantage that it has error proportional to Δt^n with only a small additional computational price.

Straightforward integration of the N coupled atomic equations of motion, using one of the above algorithms, produces a simulation with a fixed number of particles (N), a fixed volume (V) and a fixed energy (E) (= NVE ensamble). An NVE simulation therefore produces sample configurations from the micro–canonical ensemble. To produce sample configurations from different ensembles the system can be coupled to a heat bath at constant temperature (NVT or canonical ensemble) and also can be coupled to pistons at constant temperature and pressure (NPT or pressure-temperature ensemble). Such couplings are achieved by extending the phase to incorporate degrees of freedom corresponding to the scaling variables used to scale velocities and the simulation cell.

For example, the Nose–Hoover thermostat is a widely used temperature coupling scheme that produces simulations in the Canonical (NVT) ensemble. As described in textbooks (see, for example, Frenkel and Smit 2002) this is accomplished by adding an additional canonical coordinate, s, and conjugate momentum, $Q\dot{s}$, pair to the system Hamiltonian, and introducing a scaling factor to the velocity of each atom by a factor s. The system Hamiltonian then has the added term:

$$\frac{Q}{2}\dot{s}^2 + k_B T L \ln s \qquad (1.10)$$

where Q, representing an effective "mass", and L are parameters. With the choice $L = 3N + 1$, it can be shown that this additional term has the effect of coupling the system to a temperature reservoir at the temperature T, producing a canonical ensemble for the original system. The scaling factor has the effect of keeping the kinetic temperature of the overall system at the value of T input in the new term. A parallel method can be used to place the simulation in contact with a "piston", leading to an NPT ensemble simulation.

Hybrid ensembles such as constant pressure in one direction and constant surface area perpendicular to this direction (NAPT) can be constructed. An ensemble that is of use in lipid bilayer simulations with small numbers of molecules is one in which the surface tension (γ) of the bilayer is fixed along with the normal pressure on the bilayer (NγPT ensemble). This is necessary because, as shown by Feller and Pastor (1996) and discussed originally by Chiu et al. (1995), small lipid bilayer simulations run in an NPT ensemble generally do not approach the expected equilibrium state where $\gamma = 0$.

A simulated lipid bilayer is obviously very small compared to the corresponding experimental bilayer. In such a small system the boundaries can have significant effect on the physical properties of simulated systems. The usual way to reduce boundary effects is to impose periodic boundary conditions, where the system is replicated indefinitely in all the directions. However Pastor and co-workers (2002) have described alternative boundary conditions that allow for a different type of molecular exchange within a leaflet of a bilayer and across leaflets at the boundaries. This reduces the artificial periodicity that is imposed by periodic boundary conditions. For the simulation of a planar lipid bilayer embedded in a three-dimensional bath of water, the two axes in the plane of the bilayer are to be distinguished from the third dimension normal to the bilayer. If an NPT ensemble is to be used care must be taken in the implementation of the pressure coupling algorithm. Most MD code packages implement a constant pressure simulation by allowing the dimensions of the three axes to change during the simulation, in response to an applied pressure (this is implemented in several different ways that are beyond the scope of this chapter to describe). The user must choose whether to allow the three sides of the simulation box to be changed isotropically or to decouple the three sides of the box. For a bilayer the best approach is to decouple the normal dimension, but couple the two dimensions of the box parallel to the bilayer plane (this avoids unphysical changes in the shape of the membrane in the box).

1.2.3
Molecular Dynamics Simulations of Lipid Bilayers

The previous sections focused primarily on general molcular dynamics methodologies, with some examples for lipid bilayers. In this Section we now turn to the specific design and execution phases of a molcular dynamics simulation of a lipid bilayer. These include the initial state construction, equilibration of the simulation and the overall length of the simulation.

1.2.3.1 System Construction and Simulation Design

An initial salient point to consider in the design of an molcular dynamics simulation of a lipid bilayer is that single component lipid bilayers exhibit a first order phase transition with respect to temperature. This temperature ranges from about $40\,°C$ for DPPC to temperatures well below $0\,°C$ for lipids with one or more double bonds in the hydrocarbon chains (see the *LIPIDAT* data base of thermochemical properties of lipids: www.caffreylabs.ul.ie). Above the phase transition temperature lipid bilayers are in a fluid or liquid crystalline phase and it is in this region that molecular dynamics is especially suitable tool to study structural and dynamic properties. Additionally, the structure of the fluid phase varies from lipid to lipid. Phospholipids such as dioleyol phosphatidylcholine (DOPC) and palmitoyloleyol phosphatidylcholine (POPC) are relatively highly disordered above their respective phase transition temperatures, due to the presence of double bonds between carbons 9-10 on one (*sn*–2) chain (POPC) or both chains (DOPC). However, under similar conditions, sphingolipids are significantly more ordered. The key difference in structure between phospho-sphingolipids (SM) with 16-carbon chains, and a saturated phospholipid with 16-carbon chains, DPPC, is small. SM has one double bond at the 4-5 position on the *sn*–2 chain, and a sphingosine backbone region that contains amide and hydroxyl moieties, compared to carbonyls in DPPC. 18:0 SM has a phase transition temperature of around $45\,°C$, close to that of DPPC. Nonetheless, the area per molecule in 18:0 SM above the phase transition is considerably smaller than that of DPPC (about $64\,\text{Å}^2$ for DPPC, and about $52\,\text{Å}^2$ for 18:0 SM). This means that simulations of SM bilayers will involve more ordered and rigid membranes, compared to phospholipids, so that configuration sampling will require longer simulation times. This is also true for mixed bilayers where one of the components is an ordered lipid, say SM or cholesterol. Interestingly if one removes the phosphocholine polar group from SM one obtains ceramide, an important skin lipid and a participant in some cell signaling phenomena. Ceramide bilayers are even more highly ordered than SM bilayers, with phase transition temperatures above $90\,°C$. Figure 1.6 shows a snapshot of a ceramide bilayer run at $95\,°C$. We will discuss simulations of mixed and ordered (but still "fluid" phase) bilayers in a subsequent section.

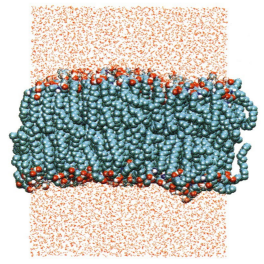

Fig. 1.6 Snapshot of a hydrated (with 50 waters/lipid) ceramide bilayer at $95\,^\circ\mathrm{C}$.

A second salient point to consider in the design of a molcular dynamics simulation is the duration of the simulation run. This is determined by the slowest degree of freedom that one wants to study. For example, in simulations of mixed lipid bilayers in which the lateral organization and distribution of the various lipids is important the simulation time is determined by the diffusion coefficient of the lipid constituents. Unfortunately for molecular dynamics simulations, lateral diffusion coefficients for the lipids in the liquid crystalline phase are small. As determined from experiments diffusion constants are of the order of $\sim 5 \times 10^{-12}\,\mathrm{m}^2/\mathrm{s}$ (Filippov et al. 2003). This value along with Einstein's relation gives a root mean square (RMS) displacement of $\sim 20\,\text{Å}$ for a single lipid in $200\,\mathrm{ns}$. So, if we are investigating organization on the length scale of $20\,\text{Å}$ then typical simulation runs should be several hundreds of nanoseconds. On a small size linux cluster a typical molecular dynamics simulation consisting of around 200 lipids, with an adequate amount of water, can achieve about a half nanosecond per day. Hence, one needs at least a few months of wall clock simulation time to observe organization on $20\,\text{Å}$ organization.

In simulations a small patch of a bilayer self–assembles in a lipid water solution in few tens of nanoseconds (de Vries et al. 2004). However most simulations are performed starting with pre-assembled bilayers. In either case, but especially in the case of a pre-assembled initial bilayer, a sufficiently long initial simulation is needed to ensure that the equilibrated state has no memory of the initial state. Figure 1.7(a) shows one possible choice for an initial state, a bilayer in a gel-like conformation with all straight, ordered

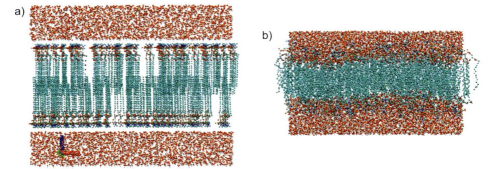

Fig. 1.7 Snapshots of sphingomyelin lipid bilayers. The lipid is 18:0 Sphingomyelin, a common brain lipid. Color coding is: gray: hydrocarbon chains; red: oxygen atoms, orange: phosphorous atoms: white: hydrogen atoms, and blue: nitrogen atoms. Water molecules and hyrdogen atoms on hyrdocarbon chains and outermost choline groups have been excluded for clarity. (a) The initial state of the bilayer consisting of all straight chains. (b) The bilayer after 20 ns of MD simulation.

chains. Figure 1.7(b) shows that same bilayer after 20 ns of molcular dynamics, in which the chains have assumed disordered, fluid-like conformations. An alternative initial state construction procedure (Venable et al. 1993; Pandit et al. 2004b) involves picking each individual chain from a large chain conformation library, and assembling a bilayer from these choices. In this method care must be taken to avoid severe steric overlaps between neighboring chains that will cause the simulation to "blow up" due to huge initial forces. A third increasingly popular option for starting an MD simulation is to simply download a pre-equilibrated lipid bilayer from one of the multiple websites that offer these data files. This approach is often employed for simulations of the effect of non-lipid molecules such as peptides, anesthetics, or channels in a bilayer environment. The non-lipid molecule is inserted into the downloaded bilayer by removal of a sufficient number of lipids to create accessible volume. The new composite bilayer is then re-equilibrated. However this latter option is useful only if the lipid of interest has already been simulated to equilibrium, and has properties that are in concordance with experiment. The simulation of the composite membrane is then subject to the details of the initial molcular dynamics simulation that created it, and should be run using the same forcefield that was used in the creation and equilibration of the downloaded bilayer. If new systems are to be simulated they should be built from the "ground up". For example, for simulations of multi-component bilayers Pandit et al. (2004b) constructed the initial bilayers of 100 18:0 SM, 100 DOPC and 100 cholesterol by randomly picking molecules from previous simulation coordinate files and placing them in a bilayer geometry with lipid phosphate locations and cholesterol oxygen locations at ± 25 Å from the origin.

In general the initial setup of the simulation state must be done with care. There should not be any artificial local or global structures in the initial state as these may produce unphysical correlations that may not disappear even after a very long simulation. Initial states with large potential energies inadvertently embedded between some of the molecules will cause major instabilities due to large repulsive forces.

In all cases where a bilayer is built it is necessary to carefully adjust the initial configurations through an energy minimization scheme, to eliminate high energy contacts present in the initial configuration. Most popular molcular dynamics simulation packages offer energy minimization algorithms using steepest descent or conjugate gradient methods (Berendsen et al. 1995). Equilibration of the lateral organization in a multicomponent membrane will, as discussed above, require many hundreds of ns. In this case an alternative is to build and run simulations with several different possible initial states. This could include different random placements of molecules or it could also include the construction of a pre-formed domain to consider possible higher degrees of lateral membrane organization (Pandit et al. 2004c). The equilibration time depends on the size and chemical composition of the simulation and on the initial state of the system. Typically equilibration is examined by calculating time correlation functions of relevant physical quantities of the system. For example, in lipid systems usually area per lipid and chain order parameters are closely monitored. Apart from physical properties, thermodynamic properties such as temperature and pressure are also closely monitored during equilibration process.

In the process of designing a molecular dynamics simulation of a lipid bilayer a decision must be made regarding the inclusion of hydrogen atoms. Simulations that include all hydrogen atoms explicitly are referred to as *all-atom simulations*. Usually all-atom simulations are more accurate but they impose a huge penalty in computation time, due to the large number of additional degrees of freedoms associated with the hydrogen atoms. Also explicit hydrogens in the system restrict the integration time step because the hydrogen atom is one order of magnitude lighter than the other atoms in the simulation. Alternatively, one can use a *united-atom simulation* where apart from hydroxyl, amide and water hydrogen atoms, all other hydrogens are combined with the atom with which they are connected. Forcefields for both types of simulations are available, as discussed in a prior section. Generally, for lipid systems the united-atom simulations are preferred because they strike a balance between accuracy and the simulation speed. However Venable et al. (2000) have shown that an all-atom model is necessary if one is to simulate a lipid bilayer in the low temperature gel phase.

To simulate a lipid bilayer in a biological context it is necessary to "hydrate" the simulation box. The user must determine the model water molecule to be used and the number of molecules to add. Here again there is a trade-off between atomistic accuracy and computational speed. Most commonly used water molecules treat water as a rigid assembly of three or four point charges (balanced for overall neutrality), (Stillinger and Rahman 1974; Stillinger 1980; Jorgensen et al. 1983). For each model the central oxygen atom also serves as a center for a 6-12 van der Waals potential. In each model locations and magnitudes of the positive and negative charges are adjusted to fit experimental data, but no model fits all of the diverse and complex properties of water. For example TIP4P (Jorgensen et al. 1983), perhaps the most popular water model used in simulations, has a dielectric constant that is much lower than 80, the value for pure water. It has a melting temperature of $232\,K$, and it fails to predict the ice 1h solid phase. The other fixed point charge models have their own shortcomings. Importantly for biological simulations, none of the rigid water molecules allow for the important effects of polarization. However, in spite of these rather severe model deficiencies, fixed-charge water models provide a degree of atomistic structural accuracy to a molecular dynamics simulation of a bilayer so that the water models may be reasonably accurate over the temperature range of molecular dynamics simulations of lipids ($10-50\,°C$). A new model for water, CC-pol (Bukowski et al. 2007), has been developed that promises to improve greatly on water molecules and includes polarization effects. The model is more complex than the above water models, however with some 3-body interactions that may adversely affect computational performance.

The number of water molecules to be included in the run is as important as the choice of water model for a simulation. In a simulation of a hydrated bilayer about 8–10 water molecules become hydrogen bonded to lipid polar group, and carbonyl oxygens of each lipid molecule. This means that the water near the lipid interface is not structurally similar to bulk water. So to simulate a lipid bilayer in excess water, it is necessary to have sufficient water so that there is a layer of water with the density and diffusion constant of bulk water. This layer serves as a buffer to interactions across the periodic boundary of the lipid polar groups on the opposite leaflets. If the leaflets interact across the periodic boundaries the simulation will contain serious unwanted artifacts. A general guideline is that there should about 32 waters per lipid for simulations of phospholipids with phosphocholine polar groups. However for lipids with more exposed charge groups, such as ceramide (Cer) more waters (in that case about 50 per lipid (Pandit et al. 2007c)) are needed, see Fig. 1.6. The hydration level is an important consideration for an additional reason, namely the computation of electrostatic interactions between water molecules utilizes much of the CPU time in a lipid bilayer molcular dynamics simulation.

Once a hydrated lipid bilayer has been successfully constructed, hydrated with the water molecule of choices and energy minimized, it must be equilibrated by molecular dynamics simulation before structural properties can be calculated. Generally equilibration is carried out by running successive simulations of a few hundred picosecond duration. After each simulation the atomic velocities are reset randomly from a Gaussian distribution with variance proportional to the simulation temperature. The number of such simulations needed to reach equilibrium depends on the specific lipid composition and the size of the simulation, but at least 10–20 ns seems to be necessary for even the simplest cases. It is also possible to include other equilibration methods, in conjunction with MD, to speed the equilibration runs. Chiu et al. (1995) found that intermittent Configurational Bias Monte Carlo steps (to be described in a subsequent section) speed up equilibration of a DPPC bilayer by a factor of three. The reason for the speed-up is that MC moves can change the configuration of the bilayer so that the MD run that follows has a different starting point, and traverses into a different region of configuration space. A simulated lipid bilayer is judged to be equilibrated after structural and thermodynamic properties of the system become independent, within fluctuations, of the time. Properties to be monitored during equilibration include the dimension of the simulation cell (in an NPD ensemble), the total energy (in NPT or NVT ensembles), and structural properties such as chain order parameters. However, we hasten to point out, that in an MD simulation run on even a 100 ns time scale not all degrees of freedom will be equilibrated. For example, molecular rotational and translational reorganization requires an order of magnitude or more longer in time.

After equilibration is achieved a long continuous molecular dynamics simulation (with no velocity resets) is run. Trajectories and snapshots are saved at regular intervals (usually a few ps) from this run for the calculation of the structural and statistical properties of the bilayer. In the next sections we will discuss the analysis steps that produce these data.

1.2.3.2 Analysis and Comparison with Experiments

Lipid bilayer systems continue to be extensively studied using various experimental techniques, and these data should be continually used to test simulations. Although atomistic simulations have a unique contribution to understanding of the properties of lipid bilayers, any comparison between simulations and reality has to be done in the context of the finite (extremely small compared to experimental) length and time scales used in simulations. In the following sections we will discuss comparison of simulations with experiments such as X-ray diffractions, NMR, and AFM experiments that determine structural properties of the lipid bilayers. We will also discuss comparison of simulations with electrophoretic measurement experiments which measure ion binding and surface potential of lipid bilayers. Agreement between simulation and any particular experimental data set is a necessary but not a sufficient condition for the validity of the simulation. Agreement between simulation and a *large set of independent experimental data sets* improves confidence in the simulations but still does not *prove* the validity of the calculations.

X-ray diffraction experiments and simulations: In experiments the form factor of the bilayer is obtained from the X-ray diffraction pattern. Then the electron density profile is obtained from the form factor by developing a lipid bilayer model and optimizing the model parameters to reproduce the observed form factors. Many different structural models have been introduced for the analysis of electron densities in lipid bilayers. Torbet and Wilkins (1976) developed a model with constant electron densities for different regions of the lipid bilayer. Due to the use of step functions the model develops discontinuities leading to high frequency noise at the large amplitudes of the form factor (see figures 5, 7, 8 of Klauda 2006). To overcome the difficulties introduced by step functions Wiener and White (1991, 1992, 1992) introduced a model in which overlapping Gaussian distributions were used. Although this model did not have discontinuities it was necessary to tune a large number of free parameters and consequently non-unique solutions were possible. Two models were developed and used by Nagle and co-workers using liquid crystal theory and, more recently, simulations (Klauda et al. 2006; Tristram-Nagle and Nagle 2004; Nagle and Tristram-Nagle 2000) to interpret X-ray diffraction data.

In simulations one can directly calculate the electron density. However, direct comparison of this electron density is not possible because the experimental electron density is model dependent. It is therefore better to use simulation data to calculate the form factors from the electron densities using

$$F(q) = \int_{-D/2}^{D/2} (\rho_e(z) - \rho_e^{\text{bulk water}}) \cos(qz) dz \qquad (1.11)$$

where D is average length of the simulation cell in z direction, $\rho_e(z)$ is symetrized electron density of the system, and $\rho_e^{\text{bulk water}}$ is the electron density of the bulk water. Figure 1.8 shows a comparison between simulations and experiment for X-ray form factors (a) and electron densities (b) for POPC.

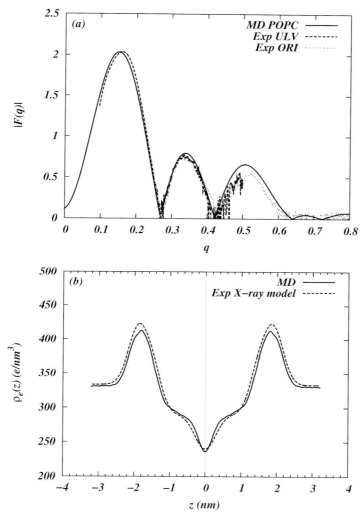

Fig. 1.8 Simulated and experimental X-ray structure data for pure POPC: (a) Form factors; (b) electron density. ULV: experimental data from unilamellar vesicles; ORI: experimental data from oriented multibilayers. Reprinted from Pandit et al. (2007a)

NMR experiments and order parameters: The ordering of hydrocarbon tails is determined in NMR experiments by measuring the deuterium order parameters. The order parameter tensor, S, is defined as

$$S_{ab} = \frac{1}{2}\langle 3\cos(\theta_a)\cos(\theta_b) - \delta_{ab}\rangle \quad a,b = x,y,z \qquad (1.12)$$

where θ_a is the angle made by the ath molecular axis with the bilayer normal and δ_{ab} is the Kronecker delta. In ^{2}H–NMR experiments quadrupolar splitting $\Delta\nu_Q$ is observed for the selectively deuterated hydrocarbon chains (Douliez et al. 1995). This splitting is related to the Wigner matrix elements $\mathcal{D}_{mn}^{(p)}$ as

$$\langle \Delta\nu_Q \rangle = \frac{3}{2} A_Q \frac{3\cos^2(\theta_L) - 1}{2} \langle \mathcal{D}_{00}^{(2)}(\Omega) \rangle \tag{1.13}$$

where θ_L is the angle between bilayer normal and the laboratory z–axis and

$$\langle \mathcal{D}_{00}^{(2)}(\Omega) \rangle = S_{CD} = \frac{1}{2} \langle 3\cos^2(\beta) - 1 \rangle \tag{1.14}$$

Figure 1.9 illustrates the definition of β; the angle between the bilayer normal and the $C-H$ plane for a given methylene molecule. In the simulations with the united atom forcefield, the order parameter for saturated and unsaturated carbons S_{CD} can be determined using the following relations

$$-S_{\mathrm{CD}}^{\mathrm{Sat}} = \frac{2}{3} S_{xx} + \frac{1}{3} S_{yy} \tag{1.15}$$

$$-S_{\mathrm{CD}}^{\mathrm{Unsat}} = \frac{1}{4} S_{zz} + \frac{3}{4} S_{yy} \mp \frac{\sqrt{3}}{2} S_{yz} \tag{1.16}$$

These relations essentially determine β using tetrahedral and planer geometry of saturated and unsaturated bonds respectively. Figure 1.9 shows an order parameter profile for DPPC bilayer at $50\,°C$. A signature of the profiles is the decrease in the order parameter with increasing carbon number along the chain.

Most current simulations are capable of producing order parameter profiles that agree with experiment. That this agreement is found in spite of very short simulation times (far shorter than the time scale for whole molecule tilting or large scale rotations) suggests that simulations are able to generate enough molecular configurations to fully sample the important region of configuration space for a lipid bilayer. The slope of the profile in Fig. 1.9 is a general consequence of steric interactions between chains in a bilayer environment.

Atomic force microscopy experiments and simulations: In atomic force microscopy experiments a bilayer is generally supported on a hard surface like mica and then thickness and surface forces are deduced by scanning the bilayer surface with a cantilever.

To reliably measure the bilayer thickness in simulations consider first the simulation setup in comparison with the AFM experimental setup of Rinia et al. (2001). In the experiment the thickness is measured with respect to a flat surface on which the bilayer is supported. Since the simulated bilayer does not have such a flat reference surface an algorithm proposed by Pandit et al. (2003a) can be employed which gives a surface to point correlation function.

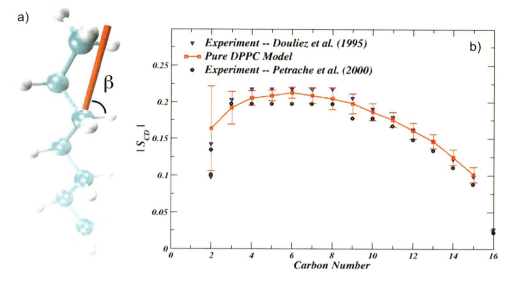

Fig. 1.9 (a) Schematic diagram illustrating the angle used in the determination of segmental order parameters. (b) Typical plot of an order parameter profile comparing experiment and simulation

The algorithm is described in the schematic drawing in Fig. 1.10. Here, for each phosphorus in the top leaflet one locates the phosphorus in the lower leaflet which is approximately below it. This is achieved by

- Tesselating the lower leaflet into voronoi polygons.
- Projecting coordinates of phosphorus from the top leaflet on to this tesselated surface.
- Identifying the polygon in which the projected coordinates fall. This procedure identifies a trans-bilayer "neighbor" for each lipid in the top leaflet.

With such identification we define the distance of phosphorus in the top leaflet with respect to the surface defined by the phosphorus atoms in the lower leaflet as the normal distance between phosphorus atoms from two leaflets that are "vertical neighbors" of each other. This distance is used to calculate the densities of phosphorus atoms of DOPC and SM in one leaflet with respect to the surface defined by the phosphorus atoms in the other leaflet.

As an example we compute the thickness of a ordered domain in fluid phase lipid bilayer. Figure 1.11 shows plots of the densities of phosphorus atoms of SM and DOPC molecules in one leaflet as function of the distance from the surface defined by the phosphorus atoms from the other leaflet. The SM density shows two peaks. The peak at ~ 4.5 nm thickness is mainly

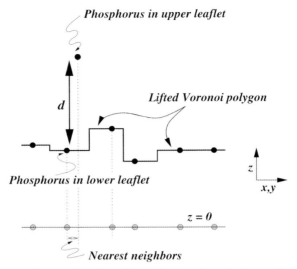

Fig. 1.10 Schematic drawing describing the method used to calculate surface to point correlation function.

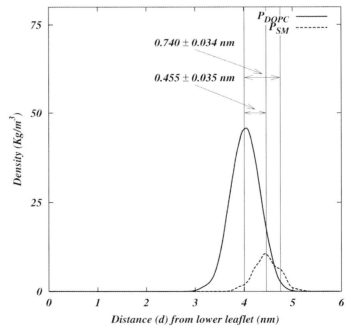

Fig. 1.11 Surface to point correlation of phosphorus in one leaflet with respect to the surface defined by the phosphorus atoms in the other leaflet. Reprinted from Pandit et al. (2004b).

due to the SM molecules which are on top of DOPC or other SM molecules with lower order parameter. The peak at ~ 4.8 nm represents SM molecules that are near the center of the domain where SM molecules lie only on top of another straight chain of SM molecules. The difference in the thickness of the SM–CHOL domain and the thickness of DOPC calculated from Fig. 1.11 is $\sim 4.5 \pm 0.35$ Å for the SM closer to the boundary and $\sim 7.4 \pm 0.34$ Å for the SM near the center of the raft like domain. The error estimates of the thickness were calculated by computing the standard deviation of the average thickness calculated over several 250 ps trajectories. Rinia et al. (2001) found this difference in AFM experiments to be ~ 6 Å.

1.2.3.3 Surface Potential Experiments

Experimental studies measuring the electrostatic properties of the membrane interface often make use of electrophoretic methods (Cevc 1990; Eisenberg et al. 1979; McLaughlin 1989; Tatulian 1987). In electrophoretic experiments the mobility of vesicles in electrolyte is measured. The ζ–potential is calculated from this mobility using the Helmholtz–Smoluchowski equation. The membrane surface charge density is then determined by calculating the intrinsic binding constant of ions using the Langmuir isotherm along with the Gouy–Chapman (GC) theory. The GC model assumes that the interface between the membrane and aqueous solution is planar with zero width and that the charge on the membrane is homogeneously distributed on the membrane surface in a continuous way. The ions in the GC description are represented as point charges immersed in a dielectric continuum and the ion–ion correlation is neglected.

To illustrate the analysis of simulation data for ion binding to a lipid bilayer we summarize the study of Pandit et al. of a DPPC bilayer in a salt solution (Pandit et al. 2003b). In order to consider ion binding it is necessary to identify a reasonable criterion to discern if an ion is bound. One of the simplest structural changes that indicates the binding of an ion is its dehydration. If an ion loses one or more of the water molecules from its coordination shell it can be considered to be bound to the membrane surface to some extent. Figure 1.12 shows that Na$^+$ ions in the range of 1.4–1.8 nm from the center of the bilayer have 1–3 fewer waters (out of ~ 6) in their coordination shell. One may consider that only these interfacial ions are bound to the surface. With this criterion $\sim$0.9–3.7 ions are bound to the surface. This range in the number of bound ions gives rise to an intrinsic binding constant[1] K_{Na^+} of $\sim$0.15–0.61 M^{-1} (Pandit et al. 2003c). The

---

1) The intrinsic binding constant is taken to be $K = \frac{\alpha}{(1-\alpha)C}$, where C is the concentration of ions at the membrane surface (in our case $\sim$0.1 M) and α is the fraction, (moles of bound ion) / (moles of lipid on the surface) (Macdonald and Seelig 1988). The number of lipids on the surface of a bilayer leaflet in the case of our simulation is 64.

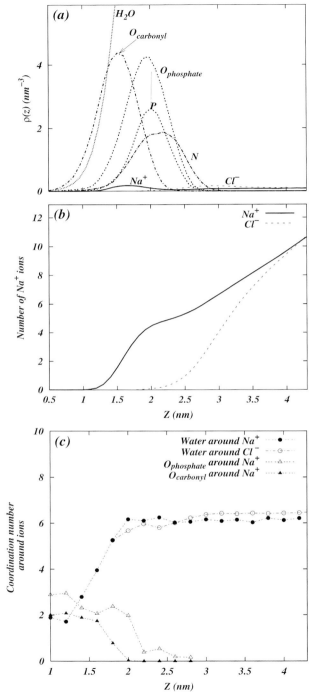

Fig. 1.12 Plots of (a) atom density, (b) sodium atom distribution and (c) coordination numbers from simulations.

experimentally observed binding constant of Na^+ to phosphatidylcholine is 0.15±0.10 M^{-1} (Tatulian 1987). Following the same logic simulations show that one Cl^- is bound to the membrane surface (Pandit et al. 2003b). This gives a binding constant K_{Cl^-} of ~0.16 M^{-1}. The corresponding experimental value is 0.2±0.1 M^{-1} (Tatulian 1987).

In order to probe the electrostatic environment at the membrane surface in a simulation one can calculate the electrostatic potential as a function of the bilayer normal (z) by twice integrating Poisson's equation for the charge density along the bilayer normal z axis as follows:

$$\Phi(z) - \Phi(z_0) = \frac{-1}{\epsilon_0} \int_{z_0}^{z} \int_{z_0}^{z'} \rho(z'') dz'' \, dz' \quad (1.17)$$

where the point z_0 is in the bulk water, ϵ_0 is the permitivity of the vacuum, and the ρ is the charge density calculated by dividing the whole box into slabs parallel to the x–y plane and counting the number of charges in each slab. The zero of the potential is placed at z_0 and, since leaflets of the bilayer are equivalent, we averaged the contributions from the two leaflets in the calculations. Figure 1.12 shows the potential profile from the center of the bilayer to the bulk water. The PC—NaCl system exhibits a positive potential (~25mV) with respect to the bulk water just outside the bilayer ($> \sim 3$ nm), while the potential in the pure-PC system remains approximately zero. We attribute this positive potential to the adsorption of sodium ions to the surface of the membrane. We calculated the surface charge density as a function of z using the following relation:

$$\sigma(z) = \int_{0}^{z} \hat{\rho}(z') dz' \quad (1.18)$$

where $z = 0$ is at the center of the bilayer and $\hat{\rho}$ is the charge density of the system excluding water. Figure 1.12 shows that after ~2.65 nm from the center of the bilayer the surface charge density becomes positive due to the adsorption of excess sodium ions in the PC—NaCl system. Note that the surface charge density in the pure-PC system is never positive. The observation of positive charge density and of a small positive potential in a region close to the boundary of the bilayer and water is consistent with the sign and the value of the ζ–potential measured in experiments with Na^+ cations (Makino et al. 1991). The experimental results related to ζ–potential are often analyzed with the help of the simple Gouy–Chapman theory.

Of general interest for lipid bilayer simulations is the dipole potential of the membrane. This is the potential difference between the interior and the surface of the membrane. Figure 1.13 shows that this difference is a few hundred millivolts. It is possible to use simulation data to identify the source of this potential. Figure 1.13 shows calculated dipole potential profiles across the membrane for 18:0 sphingomyelin and 16:0 ceramide bilayers. The sim-

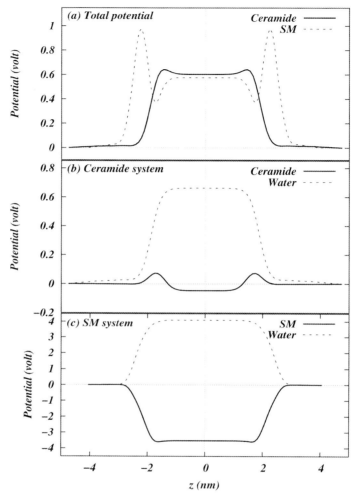

Fig. 1.13 Dipole potential profiles for SM (dashed lines) and Cer (solid lines) bilayers. (a) Total potential. (b) Separate contributions from water and lipid for Cer. (c) Separate contributions from water and lipid for SM.

ulations both show that the positive potential barrier is the result of the difference between the potentials from larger contributions: water contributes a positive dipole potential of several volts for SM and less for Cer. The fact that the net potential barrier is a few millivolts means that the lipid polar groups contribute a negative dipole potential that does not quite compensate the water contribution.

1.2.3.4 Radial Distribution Functions

While snapshots of configurations produced in lipid bilayer MD simulations are interesting, quantitative structural analysis of intermolecular and interatomic structures in a fluid bilayer requires the calculation of radial distribution functions. If properly sampled the calculated correlation functions can provide information about change in free energy with the configuration of two particles as the reaction coordinate. Generally pair correlation functions are calculated as averages over solid angles leaving only a radial dependence. These are denoted as *radial distribution functions* (RDF). The RDF is defined as

$$g(r) = \frac{N(r)}{4\pi r^2 \rho \delta r} \quad (1.19)$$

where $N(r)$ is the number of atoms in the shell between r and $r + \delta r$ around the central atoms and ρ is the number density of atoms, taken as the ratio of the number of atoms to the volume of the simulation cell. In a simulation the RDF is calculated, between atoms A and B, on different molecules by scanning all simulation snapshots from the equilibrated bilayer simulation, calculating the radial distance between each pair A and B and binning the data. The normalized binned distributions then represent the RDF. For correlation between complex polyatomic molecules it is necessary to pick specific atoms in each molecule for binning. For example, a typical RDF produced from a simulation is shown in Fig. 1.14 in which atom A is the cholesterol hydroxyl oxygen and atom B is a water oxygen atom.

The RDF in this figure was produced from an MD simulation of a ternary 1 : 1 : 1 mixture of DOPC, SM and cholesterol. The RDF distances were calculated between the hydroxyl oxygens of cholesterol molecules and the middle carbon atom of the backbone regions of DOPC and SM, respectively. It is important to point out that the RDFs are quite sensitive to the choice of atoms. For example, if RDFs between lipids and cholesterol are calculated based on center of mass distances much structural information is hidden because the center of mass distances are strongly dependent on molecular conformations that are driven by the hydrocarbon tails. Lateral organizational information is better determined from RDFs calculated between the backbone regions, to which all of the chains are attached.

Peaks in Fig. 1.14 are identified as coordination shells for the molecules. The first coordination shell is defined by the location of the sharp peak in the RDF. By integrating the RDF under the peak one can obtain the total number of waters bound to cholesterol. RDF data as defined above are purely radial. However in the case of bio-molecules angular asymmetry is also important. The cholesterol molecule has one flat face (the α–face) and one face which is rough due to protruding methyl groups (the β–face). Since cholesterol lies primarily in the hydrocarbon region of the bilayer it is reasonable to ques-

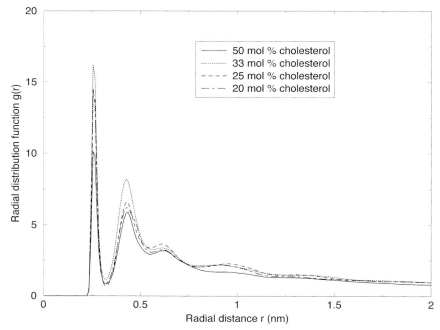

Fig. 1.14 RDF of cholesterol OH – water O in DPPC at various cholesterol contcentrations. Reprintred from Chiu et al. (2002).

tion whether this specific design of the cholesterol molecule plays any role in promoting domain formation. For correlations involving two variables, orientation and position, one can define a bivariate correlation function $g(r, \varphi)$ between one selected backbone carbon atom of DOPC and SM molecules respectively, and the oxygen atom of cholesterol defined by

$$g(r, \varphi) = \frac{N(r, \varphi)}{2\pi r \rho \delta r \delta \varphi} \qquad (1.20)$$

where the distance r and ρ are defined as in RDFs, the angle φ is the angle made by the distance vector with respect to the positive x–axis of the cholesterol body coordinate frame (see Fig. 1.15) and $N(r, \varphi)$ is the number of the selected lipid carbon atoms in an area element $r\delta r\delta\varphi$ at the point (r, φ) from the oxygen of cholesterol. In a later section we will illustrate an example of the use of a two dimensional RDF that has provided new insights into lipid-cholesterol interactions, and has led to an accurate coarse grained model for lipid-cholesterol bilayers.

1.2.3.5 Heterogeneous Membrane Simulations

In this section we describe results of simulations of heterogeneous membranes. The first and simplest type of heterogeneous membrane that has

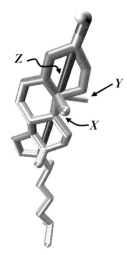

Fig. 1.15 The body coordinate system used in calculation of $g(r, \varphi)$. Reprinted from Pandit et al. (2004b).

been studied by simulations is also one with considerable biological importance: bilayers composed of a single species of lipid and a varying concentration of cholesterol. Lipid-cholesterol mixtures have been studied by Monte Carlo and MD simulation for over twenty years by several groups (Scott 1993; Tu et al. 1998; Pandit et al. 2004a; Hofsäß et al. 2003; Chiu et al. 2001a; Chiu et al. 2001b; Scott 2002). In 2002 Chiu et al. (2002) reported a comprehensive study of mixed bilayers of dipalmitoyl phosphatidylcholine (DPPC) and cholesterol. They carried out Configurational Bias Monte Carlo and molecular dynamics simulations for bilayers of dipalmitoylphosphatidylcholine (DPPC) and cholesterol for DPPC:cholesterol ratios of $24:1$, $47:3$, $23:2$, $8:1$, $7:1$, $4:1$, $3:1$, $2:1$ and $1:1$, using 5 nanosecond (ns) molecular dynamics runs and interspersed Configurational Bias Monte Carlo to ensure equilibration (this procedure will be described in a subsequent section). For simulations with cholesterol concentrations above 12.5% the area per molecule of the heterogeneous membrane varied linearly with cholesterol fraction, as shown in Fig. 1.16. From the slope of the linear area versus cholesterol concentration it was found that the area per cholesterol was surprisingly small, $\approx 24\,\text{Å}^2$ per molecule, while the area per DPPC was $\approx 52\,\text{Å}^2$. The low area of cholesterol is a consequence of the ability of the flexible DPPC chains to back closely around the cholesterol rings. Figure 1.16 shows the linear relationship between area and cholesterol concentration.

Radial distribution function analysis of the lateral distribution of cholesterol molecules in the bilayer revealed a tendency for small subunits of one

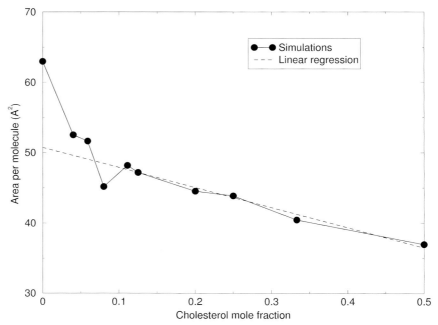

Fig. 1.16 Plot of area per molecule vs cholesterol concentration for DPPC-cholesterol simulations. Reprinted from Chiu et al. (2002).

or two lipids plus one cholesterol, hydrogen bonded together, to act as one composite particle and perhaps to aggregate with other composites at higher concentrations of cholesterol. The conclusions drawn from simulations are consistent with experimentally observed effects of cholesterol, including the condensation effect of cholesterol in phospholipid monolayers and the tendency of cholesterol-rich domains to form in cholesterol-lipid bilayers, but the short simulation times shed no light on the longer-time stability of the subunits. However the data shown in Fig. 1.16 and the insights that resulted from the DPPC-cholesterol simulations are an early example of the powerful insight that can come from careful simulations.

In an extension of the work of Chiu et al. (2002), and with the revised and improved forcefields for lipids described in the previous section Pandit et al. have run and analyzed simulations of mixtures of cholesterol with palmitoylolyeol phosphatidylcholine (POPC) and diolyeol phosphatidylcholine (DOPC) (Pandit et al. 2007b). This work revealed a new way to interpret the area per molecule in a pure lipid bilayer, as discussed in the previous section. It also revealed significant new insights into the nature of phospholipid-cholesterol interactions.

Table 1.2 Partial molecular volumes and areas calculated for lipid-cholesterol mixtures calculated from simulations and from experimental data (Greenwood et al. 2006).

	DPPC (MD)	DPPC (Expt)	POPC (MD)	POPC (Expt)	DOPC (MD)	DOPC (Expt)
Temperature (K)	323	323	303	303	303	303
V_l (Å^3)	1213.9 ($x < 0.25$)	1228.6 ($x < 0.25$)	1239.5	1255.5	1269.4	1300.1
	1197.3 ($x \geq 0.25$)	1207.6 ($x \geq 0.25$)				
V_{ch} (Å^3)	594.4 ($x < 0.25$)	573.8 ($x < 0.25$)	608.0	622.6	606.8	632.9
	644.9 ($x \geq 0.25$)	637.1 ($x < 0.25$)				
A_l (Å^2)	54.07	—	61.50	—	61.50	—
A_{ch} (Å^2)	23.84	—	12.42	—	20.84	—

x is the cholesterol concentration. Partial specific area data are for $x > 15\%$.

Table 1.2 lists calculated partial molecular volumes and areas calculated from the simulations for DPPC, DOPC and POPC as functions of cholesterol concentration. Pandit et al. calculated the partial molecular following the method proposed by Edholm and Nagle (2005), and the partial specific volume following the method proposed by Greenwood et al. (2006), and described earlier in this chapter. Of particular interest are the partial molar molecular areas. For cholesterol concentrations close to 10% the partial molecular area of a cholesterol is negative. Negative partial molecular areas are interpreted by Edholm and Nagle (2005) as a manifestation of the condensation effect of cholesterol on surrounding lipids. Above about 15%, and up to 50% cholesterol concentration, the partial specific areas take on the concentration-independent values listed in Table 1.2. The partial molecular areas of cholesterol and lipids are listed in Table 1.2. The computed cholesterol partial molecular area is largest in saturated DPPC and smallest in POPC where one chain is saturated and the other chain is unsaturated. For DOPC, with two unsaturated chains, the area per cholesterol is higher. This implies that in mono-unsaturated POPC bilayers cholesterol admits a different packing structure as compared to the mixtures with fully saturated DPPC and di-unsaturated DOPC.

To further explore this somewhat surprising conclusion, Pandit et al. calculated radial distribution functions of CH=CH group of sn–2 chains around the CH$_3$ group of β–face of cholesterol. Figure 1.17(a) shows these RDFs for POPC and DOPC molecules. There are two peak in the RDFs. The first peak is around ~ 4.5 Å from the methyl group and the second peak is around ~ 7 Å from the methyl group. As illustrated in Fig. 1.17 these two peaks

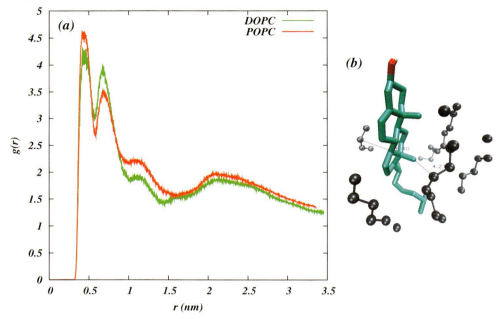

Fig. 1.17 (a) Radial distribution functions between cholesterol methyl groups on the rough β face and POPC and DOPC CH=CH double bonds. (b) Stick model of cholesterol and DOPC illustrating atoms used in RDF calculation.

correspond to two closest positions for CH=CH group along β and α–faces respectively. In Fig. 1.17, for DOPC both the peaks are approximately the same height whereas for POPC the first peak is taller than the second peak. This implies higher correlation of POPC CH=CH groups with the β–face of cholesterol.

Sphingomyelin is an important component of the outer leaflet of mammalian cell membranes, yet much less is known about SM structure in bilayers than is known about phospholipids. To begin to address this issue Chiu and co-workers constructed a fully hydrated bilayer of 18:0 SM (Chiu et al. 2003). The size of this system was 1600 SM molecules and 51 200 water molecules, for a total of over 250 000 atoms. Chiu et al. found that the SM bilayer is much more ordered than its DPPC counterpart, with an area per molecule of $48\,\text{Å}^2$ compared to about $64\,\text{Å}^2$ for DPPC. For the SM bilayer there is significant intramolecular hydrogen bonding between the phosphate oxygen and the amide hydrogen, as well as between water and the sphingosine hydroxyl moiety.

Khelashvili and Scott (2005) ran simulations of a bilayer consisting of 266 SM, 134 cholesterol molecules and 12924 waters, at two separate temperatures, $20\,^\circ\text{C}$ and $50\,^\circ\text{C}$. This choice of temperatures brackets the SM phase transition temperature of about 41 °C so that the effect of cholesterol on both gel-like and fluid-like SM bilayers would be compared. The simulations revealed an overall similarity of both systems, despite the 30 °C temperature difference. The area per molecule, lipid chain order parameter profiles, atom distributions and electron density profiles are all very similar for the two simulated systems. Khelashvili and Scott also observed strong intra-molecular hydrogen bonding in SM molecules between the phosphate ester oxygen and the hydroxyl hydrogen atoms and they found that cholesterol hydroxyl groups tend to form hydrogen bonds primarily with SM carbonyl, methyl and amide moieties and to a lesser extent methyl and hydroxyl oxygens.

Ternary mixtures of lipids and cholesterol have been studied in simulation by Pandit et al. (2004c, 2004b) and most recently by Niemală et al. (2007). In two sets of simulations Pandit et al. examined the structure of lipid bilayers containing mixtures of 18 : 0 SM, DOPC and cholesterol. In the first case they constructed a large bilayer consisting of 1424 molecules of DOPC, 266 molecules of 18 : 0 SM, 122 molecules of cholesterol and 62,561 water molecules. Figure 1.18 shows snapshots of this bilayer.

A second ternary mixture simulation run by Pandit et al. consisted of 100 DOPC, 100 SM and 100 cholesterol molecules plus 9600 waters. Unlike the domain simulation described above this system was started from a random distribution of DOPC, SM and cholesterol. They simulated this system for 250 ns to identify the structural and dynamical parameters that drive the formation of domains. Also unlike the domain simulation described above this simulation was started from a random distribution of DOPC, SM and cholesterol molecules. As a control a simulation of a binary system consisting of 100 SM plus 100 DOPC, with no cholesterol, was also run. Figure 1.18 shows before and after snapshots of the ternary system and a 200 ns snapshot of the binary system.

One of the features qualitatively revealed in the snapshots in Fig. 1.18 is that in 1 : 1 : 1 DOPC-SM-cholesterol mixtures an apparent preponderance of cholesterol exists at the interface between regions rich in SM and regions rich in DOPC. To quantitatively investigate this observation we calculated two-dimensional RDFs as described in a previous section. Figure 1.19 shows the two dimensional RDFs for DOPC-cholesterol (a) and SM-cholesterol (b). The density plots show that SM is preferentially located on the smooth face of cholesterol. DOPC tends to be found to a greater extent near the rough face and the edges of the cholesterols. The two-dimensional RDFs shown in Fig. 1.19 show how SM is found to a greater extent at the smooth face of cholesterol, while DOPC by "default" is found to a slightly greater extent near the rough face.

This result is an example of how the atomic level of details available in a simulation can reveal new subtle intermolecular interactions and correla-

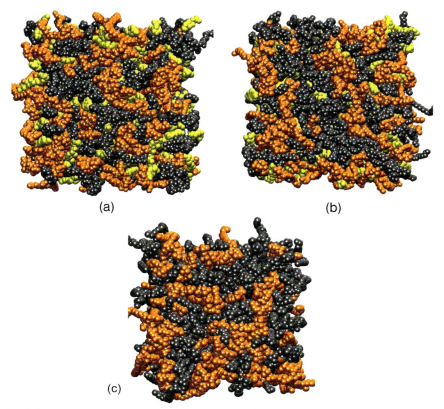

Fig. 1.18 Initial (a) and final (b) snapshots of one of the leaflets in systems of 1 : 1 DOPC-SM, with (a) and (b) and without (c) equimolar cholesterol. DOPC molecules are represented by gray solid atoms, SM by orange solid atoms, and Chol by yellow solid atoms.

tions that may be important for larger scale membrane structure. In fact, as we describe in a subsequent section, this particular find was incorporated into a coarse grained model for lipid-cholesterol mixtures that agrees extremely well with experimental data.

Niemela et al. (2007) extended the time and length scales of lipid-cholesterol simulations by running MD studies of 1024 lipids in 1 : 1 : 1 and 2 : 1 : 1 POPC:16:0 SM:cholesterol. The simulations were run for 100 ns each. This study provides pressure profiles and order profiles that show the rigidity of the bilayers, possibly a consequence of the use of POPC instead of DOPC as the phospholipid. The lateral distribution of chain order is indicative of the beginnings of a lateral organization process. These profiles are similar to ones that can be constructed by Self Consistent Mean Field modeling as we describe in a subsequent section.

1.2 Atomistic Models | 43

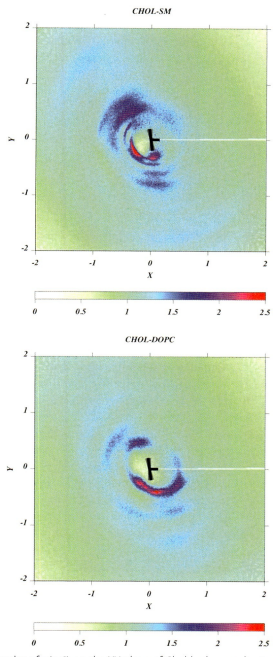

Fig. 1.19 Density plot of g(r, θ) on the XY-plane of Chol body co-ordinate system. A Chol molecule is schematically shown as a black bar with a the small appendage that denotes the β-face of the molecule with possible fluctuation of its shape (with respect to the body coordinates) shown in gray. Reprinted from Pandit et al. (2004b).

In summary, a number of important insights into interactions between lipids and cholesterol:

- Simulations have observed the change from fluid to liquid ordered structure in DPPC as cholesterol concentration increases.

- Simulations have shown that the partial specific area of cholesterol in DPPC is less in all cases than the area of cholesterol in crystals of pure cholesterol. POPC bilayers minimize the partial specific area of cholesterol.

- The asymmetric structure of cholesterol plays an important role in lipid-cholesterol interactions, with the smooth face favoring contact with saturated lipid chains strongly over unsaturated chains.

- Cholesterol seems to prefer to lie at an interface between lipids with saturated chains and lipids with monounsaturated chains or, in the case of POPC, seems to invite an asymmetric packing that minimizes the cholesterol partial specific area.

1.2.3.6 Simulations of Ordered Lipid Phases

At temperatures below the main lipid chain melting temperature lipids in bilayers form a "gel" phase in which hydrocarbon chains are generally oriented parallel to each other and tilted perpendicular relative to the membrane plane. In addition certain double chain lipids with relatively large polar groups (such as phosphatidylcholine phospholipids) exhibit an intermediate phase between the fluid and gel phases in which the bilayer forms periodic corrugations, or "ripples". While the gel phase and the ripple phase are not of direct biological significance they are of interest as new phases of soft matter. Simulations of ordered phases pose a new set of problems, because of the extra rigidity and slower rate at which configuration space is sampled. Thus equilibration is a much more severe problem in the simulation of ordered or partially ordered membranes. The slow rate of conformational changes and diffusion means that the relaxation of a simulation from the initial state means that memory of that initial state persists for much longer simulation times than would be the case for a fluid phase simulation.

A careful simulation of a DPPC bilayer in the gel phase (at $293\,\mathrm{K}$) was carried out by Venable et al. (2000). They were able to reproduce many experimental properties of the gel phase including D-spacing, chain tilt, fraction gauche and the lateral compressibility modulus. Their simulations reproduced the correct orientation of the chain tilt across the two leaflets. The tilt of the chains should result in a parallel alignment of the chains across the two leaflets. However, earlier simulations (Scott and Clark 1996, unpublished) led to herringbone patterns for the chains across the leaflets. Simulations by Tu et al. (1996) also found a herringbone gel structure. These results indicated that either the forcefields or the simulation starting state were in need of modification. Venable et al. used an all-atom model for DPPC, with all hydrogens included, whereas our own earlier simulations were done in a united atom model. It may be that all-atom simulations are needed to correctly model the gel phase in atomistic simulations.

The ripple phase has been an enigma for modeling for many years (Scott 1984). This phase is only found in lipids with relatively large polar groups and saturated or monounsaturated chains such as DPPC and POPC. It is not found in phosphatidylethanolamine bilayers. The distinguishing characteristic of the ripple phase is a periodic but asymmetric corrugation that can persist defect free for several hundred nanometers.

Numerous experimental studies have provided detailed data for the structure of the ripple phase using X-ray scattering and scanning microscopy (Stamatoff et al. 1982; Wack and Webb 1988; Sun et al. 1996; Woodward and Zasadzinski 1997; Sengupta et al. 2003). The consensus is that the ripple phase consists of asymmetric linear parallel corrugations across the membrane. The ripple wavelength varies for different lipids but is generally between 11 and 20 nm. The asymmetry presents itself in the form of major and minor domains. The major domain has structure similar to the gel phase of the lipid under study while the minor domain shows significantly greater disorder and possible chain interdigitation.

Early models of the ripple phase were based on coarse grained Statistical Mechanics models (Doniach 1979; McCullough and Scott 1990; Carlson and Sethna 1987). These models attempted to include the frustration in molecular packing caused by the large cross sectional area of the polar group relative to that of the two hydrocarbon chains. Generally all models were able to produce a periodic ripple-like patterns but none were able to describe the asymmetric ripple structure seen in experiments. In a large scale MD simulation de Vries et al. (2005) have however observed a corrugated structure that has most of the experimental properties of the ripple phase. Figure 1.20 shows a snapshot of the ripple phase in DMPC from the simulation. The simulation suggests that the ripple phase forms as a result of packing frustration between fluid and gel phases. The minor domain is predominantly fluid and the major domain is predominantly gel. One caveat is that the simulation was run with periodic boundary conditions and shows only one full ripple, so that the periodicity of the ripples is not guaranteed. Thus for a truly definitive study it will be necessary to run a simulation at least twice as large.

1.2.3.7 Simulations of Asymmetric Lipid Bilayers

Biological membranes have inner and outer leaflets that differ significantly in structure. Outer leaflets are generally rich in sphingolipids and phos-

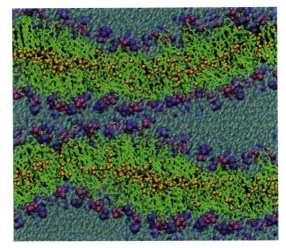

Fig. 1.20 Structure of the ripple phase from MD simulation. Note the asymmetric structure of the ripples. The simulation contains one full ripple; the figure shows periodic reflections in addition to the main simulation cell. Reprinted from de Vries et al. (2005).

phatidylcholine lipids, while inner leaflet lipids are rich in phosphatidylethanolamine and phosphatidylserine lipids. These two classes of lipids are characterized by smaller polar groups, with the bulky choline ($N(CH_3)_3$) moiety being replaced by the simpler NH_3. In addition phosphatidylserines carry a net negative charge at neutral pH.

To form a foundation for simulations of asymmetric lipids that mimic the inner and outer leaflets of membranes several groups have run simulation studies of symmetric bilayers composed of inner leaflet lipids such as ethanolamine and serine lipids. Early simulations of dilauryl phosphatidylethanolamine (DLPE) were carried out by Damodoran and Merz (1994) and by Zhou and Schulten (1995). These were very short simulations for small systems, limited by computing power in the mid 1990s. Murzyn et al. (2005) have recently done a 25 ns simulation of a mixed palmitoyloleoyl phosphatidylethanolamine (POPE) – palmitoyloleoyl phosphatidylglycerol (POPG) consisting of 54 POPE and 18 POPG, plus 1955 waters. The aim of this simulation was to determine the properties of a model for the inner bacterial membrane. They carried out a detailed analysis of hydrogen bonding between POPE pairs, and POPE and water, including water bridges between POPE. POPG did not appear to participate in H-bonding to the extent found with POPE. Simulations of phosphatidylserine bilayers were carried out by Pandit and Berkowitz and by Mukhopadhyay et al. (2002, 2004). Pandit and Berkowitz simulated DPPS in a bilayer with Na^+ counterions. They found an area per molecule of about $54 Å^2$, in the range of experimental values for this lipid. They found that the low area per molecule for a liquid crystalline phase is due to extensive counterion-facilitated hydrogen bonding in the polar region. They concluded that the counterions screen the negative charge on the serine, and that the remainder of the molecule behaves much like an ethanolamine bilayer (Pandit and Berkowitz 2002). Mukhopadhyay et al. (2004) did two simulations of a palmitoyloleoyl phosphatidylserine (POPS) bilayer, with Na^+ counterions and with NaCl. The goal of these simulations was to consider the effect of salt on the bilayer structure. Both simulations converged to an area per molecule of about 55 $Å^2$. The Na^+ ions displaced water from the ester region of the lipids, and the amine group participated in hydrogen bonds with phosphate and carboxylate groups on neighboring molecules.

Phosphatidylinositol represents a class of phospholipids which, while present only in small concentrations in eukaryotic cell membranes (and some bacteria), play important roles in a variety of biological functions. Phosphatidylinositol lipids (PI) are phosphatidic acids linked through the phosphate group to inositol, that is hexahydroxy-cyclohexane. In PI the presence of multiple hydroxyl groups on the cyclohexane ring allow for a large variety of binding and other interaction possibilities, and thus for a wide range of biological functions. The most common inositol lipid in animal and plant membranes is *sn*-1-stearoyl-*sn*-2-arachidonoyl-glycerophosphorylinositol 4,5-biphosphate(PIP_2). This lipid has two very different hydrocarbon chains; one saturated stearoyl chain and one poly-unsaturated arachidonyl chain. It also has two phosphates bound at the 4 and 5 carbons of the inositol ring. PIP_2 is found in concentrations up to 10% in brain lipid (Christy 2003), in blood platelets (Bodin et al. 2005) and in lesser concentrations other tissues. To date no atomistic simulations have attempted to include any of the PI lipids, in spite of the biological importance of these molecules.

One of the challenges in building and running MD simulations of asymmetric bilayers is that the asymmetry in lipid structure will lead to mechanical stresses in the membrane. This is in fact an underlying cause of the curvature of biological membranes, but in a simulation with periodic boundary conditions unwanted instabilities may arise. For this reason it is important that the two leaflets in an asymmetric simulation be constructed so that they will have the same molecular area at the simulation temperature. This means that the number of lipids in each leaflet may be different. Another way that the stress between leaflets can be moderated is through the addition or removal of cholesterol. Cholesterol, as discussed earlier in this chapter, modifies the molecular area of a bilayer by inducing increased order in the lipid chains.

To date two sets of simulations of asymmetric lipid bilayers have been published. Cascales et al. (2006) simulated a bilayer in which one leaflet contained 96 DPPC and 48 DPPS$^-$ lipids plus 48 Na^+ counterions. The other leaflet contained 120 DPPC molecules. They compared structural properties of the DPPC molecules on both leaflets to ascertain if the structure of the pure DPPC leaflet was perturbed by the other leaflet of mixed composition. They found that the lipid order parameters, polar group orientation and lateral diffusion rates of the DPPC were not perturbed by the presence of DPPS$^-$ in the opposite leaflet. However the DPPC in the same leaflet with DPPS$^-$ were, as one would expect, altered in their structure. Additionally the dipole potential of the asymmetric membrane showed an asymmetric is not symmetric in shape, with the potential between the water interface and the membrane center about 10 millivolts larger on the DPPC-DPPS side of the bilayer.

Bhide et al. (2007) simulated an asymmetric bilayer in which one leaflet contained 84 18:0 sphingomyelin and 44 cholesterol molecules, while the opposing leaflet contained 84 molecules of steroyloleoyl phosphatidyl serine (SOPS) and 44 cholesterol molecules. The asymmetric bilayer was built after separate symmetric bilayers of SM–cholesterol and SOPC–cholesterol were constructed and equilibrated. Then leaflets from each of the symmetric bilayers were joined to make an asymmetric bilayer. Figure 1.21 shows a snapshot of the asymmetric bilayer after 50 ns of MD simulation. In 50 ns of MD, Bhide et al. found that the structures of the two leaflets do not differ significantly from structures found in their simulations of symmetric bilayers of SOPC–cholesterol and SM–cholesterol. The simulation of Cascales et al. and of Bhide et al. show that mechanically stable asymmetric bilayers of 200 lipids can be constructed and simulated over a 50–100 ns timescale. In both cases the two lealets of the bilayers seemed to behave independently of each other. In biomembranes the two leaflets must interact but it is likely that much larger bilayers and longer simulations will be necessary to study these interaction modes.

1.2.4
Equilibrium Monte Carlo Methods

Motivated by emerging computational technologies, and by Cold War nuclear arms concerns, the Monte Carlo method was developed in the late 1940s at Los Alamos National Laboratory (Metropolis et al. 1953). Over the past 60 years the Monte Carlo (MC) method has become an essential tool in an extemely wide range of applications. The Wikipedia (http://en.wikipedia.org/wiki/Monte_Carlo_simulation) entry for Monte Carlo lists applications ranging from quantum chromodynamics to finance. In the condensed matter physics of complex systems, where direct analytical calculations are unfeasible, MC has proven to be an essential tool. The foundation for the MC

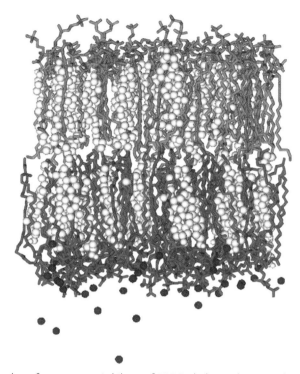

Fig. 1.21 Snapshot of an asymmetric bilayer of SOPS, cholesterol (open spheres), and Na$^+$ (solid spheres) in the lower leaflet; SM and cholesterol in the upper leaflet. Reprinted from Bhide et al. (2007).

method is described in many books and reviews, such as the classic volume of Binder (1986) and more recent texts such as that of Frenkel and Smit (2002). In this chapter we provide a synopsis of the MC method as it is used in statistical physics and we describe applications that are germane to the simulation of membranes.

The MC method provides a means for the evaluation of the equilibrium values of thermodynamic properties of many-particle systems by numerically calculating averages of the form

$$<O> = \sum_C O(C) \exp[-\mathcal{H}(C)/k_B T] / \sum_C \exp[-\mathcal{H}(C)/k_B T] \quad (1.21)$$

where the sum runs over all configurations of the system C, $\mathcal{H}$ is the Hamiltonian function evaluated for configuration C, k_B is Boltzmann's constant, T is the absolute temperature, and O is any dynamical variable. The numerator and the denominator of this equation cannot be evaluated analytically for more than a handful of nontrivial models (Baxter 1982), and these generally involve formidable mathematical analysis. Direct numerical evaluation of the sums is also not practical in many cases. To illustrate, consider a system of N n-mer linear polymers and use a model for the polymer for which only three internal torsional degrees of freedom per C–C bond are allowed. For this simplified model there are still a total of $(3^n)^N$ torsional configurations, where $N \approx 10^{23}$. The vast majority of the configurations contribute negligibly to Eq. (1.1). Many of the torsional states will be forbidden by actual or near-excluded volume overlaps. The goal in a numerical simulation is to identify and include primarily the "important" terms while wasting little time sampling states of high Boltzmann weight. If the MC summations generate trial states that have weights proportional to $\exp[-\mathcal{H}(C)/k_B T]$, then the average of a dynamical variable $<O>$ is just an arithmatic average over generated configurations,

$$<O> = \frac{\sum_{C_t} O(C_t) P_t^{-1} \exp[-\mathcal{H}(C_t)/k_B T]}{\sum_{C_t} P_t^{-1} \exp[-\mathcal{H}(C_t)/k_B T]} = \frac{1}{N} \sum_{C_t} O(C_t) \quad (1.22)$$

where C_t is a trial configuration picked with a probability given by P_t (Binder 1986). In Eq. (1.2) the terms P_t^{-1} correct for the sampling bias.

The Importance Sampling procedure was developed by Metropolis and coworkers (1953) and was designed to accomplish Boltzmann weighted sampling for simulation of many particle systems. The procedure requires that configurations be generated as a Markov walk, a succession of states for which state $n+1$ depends only upon state n. If $P(C_t)$ is the probability that trial configuration C_t occurs, and if $w(t, t')$ is the transition probability from a state t to a state t' in the Markov walk, then the rate of change of $P(C_t)$ is given by the Master Equation,

$$\frac{dP(C_t)}{dt} = \sum_{t' \neq t} [w(t', t) P(C_{t'}) - w(t, t') P(C_t)] \quad (1.23)$$

At equilibrium $\frac{dP(C_t)}{dt} = 0$ and $P(C_t) = P_{eq}(C_t) = Z^{-1} \exp[-\mathcal{H}(C)/k_B T]$ satisfies the Detailed Balance criterion:

$$P_e(C_{t'}) w(t', t) = P_e(C_t) w(t, t') \quad (1.24)$$

Utilizing the fact that the equilibrium probabilities are proportional to Boltzmann factors,

$$\frac{w(t, t')}{w(t', t)} = \frac{P_e(C_t')}{P_e(C_t)} = \exp[-(\mathcal{H}(C_t') - \mathcal{H}(C_t))/k_B T] \quad (1.25)$$

The unknown proportionality factors in the equilibrium probabilities are canceled in the ratio. The key result is that the quotient of the transition probabilities depends only on the difference in energy between the two states. It follows that a sufficient condition for the sampling probabilities to approach their equilibrium values is that the sampling procedure must be a Markov walk through phase space with step-to-step transition probabilities which satisfy Eqs. (1.24) and (1.25). Two typical examples of such probabilities are:

$$w(t, t') = \begin{array}{ll} \frac{1}{m} \exp[-(\Delta \mathcal{H})/k_B T] & \Delta \mathcal{H} > 0 \\ \frac{1}{m} & \Delta \mathcal{H} \leq 0 \end{array} \quad (1.26)$$

or

$$w(t, t') = \frac{1}{2m} \frac{\exp[-(\Delta \mathcal{H})/k_B T]}{[1 + \exp{-(\Delta \mathcal{H})/k_B T}]} \quad (1.27)$$

where m is a factor related to the number of ways to pick a succeeding move in the Markov walk and $\Delta \mathcal{H} = (\mathcal{H}(C_{t'}) - \mathcal{H}(C_t))$. More complete discussions of the underlying basis for the Metropolis MC method can be found in textbooks (Binder 1986; Frenkel and Smit 2002)

1.2.5
Monte Carlo Studies of Lipids

The Monte Carlo method, when applied in practice, consists of generating trial molecular configurations and accepting or rejecting them according to Eq. (1.26) or Eq. (1.27). Lipids, unfortunately, have very many conformational states and this means the level of sampling is reduced unless many steps are generated. For example, DPPC consists of 50 atoms (not counting hydrogens) on the three distinct chains described earlier. All three of the chains in lipids are flexible. Distortions in bond length, bond angle and, most importantly, rotations about bonds all control the conformation of the molecule.

Historically, the application of the MC method to lipids probably began with the lattice-based simulations of Whittington and Chapman (1966). Whittington and Chapman used a two-dimensional lattice to generate a line of chains using a self-avoiding walk algorithm. In general lattice-based simulations have the advantage that the configuration space of the system is greatly reduced, making long simulations of larger systems possible. The problem is that the reduction is a very severe limitation on the accessible molecular conformations in the case of large, flexible molecules such as phospholipids. As computing power has grown the need for an underlying lattice in MC simulations has diminished, and it is mainly used for models which aim to describe long range cooperative phenomena.

Early Monte Carlo continuum simulations of hydrocarbon chains in a lipid bilayer type of environment were carried out by Scott (1977). The first such simulations consisted of ten or fewer chains of hard spheres with the topmost sphere attached to an interface. Since models studied by the MC method have progressed to bilayers of 200 chains interacting via 6-12 potentials between atoms and containing cholesterol (Scott 1993) or gramicidin (Xing and Scott 1992).

In the early 1980 s, in order to carry out simulations in reasonable time it was necessary to restrict the dihedral angles allowed for $C-C$ bonds to three states per bond, 0 and ± 120 degrees. This restriction is consistent with the locations of three distinct minima in the dihedral energy function (Ryckaert and Bellemans 1978), and is commonly called the rotational isomeric approximation, and the states are called rotational isomeric states or RIS. Even within the rotational isomeric approximation it was not possible to include, for example, lipid head groups or water in MC simulations at that time. A consequence of the restricted dihedral angles was that order parameter plots for the chains in the simulations were not as smooth as experimental plots. The key lesson from these simulations was that the configuration space of a lipid bilayer with 15% or more cholesterol present is very difficult to efficiently sample by standard MC means within the rotational-isomeric model. Similar conclusions were drawn in Monte Carlo simulations of lipid chains adjacent to hard cylinders with hemispherical "lumps"(Coe and Scott 1992), and of lipid chains adjacent to gramicidin A. Comparison of data for chains of length 14, 16 and 18 carbons suggested that the hydrophobic length of the gramicidin allowed the shorter chains to pack more efficiently around it than was the case for the long chains (Xing and Scott 1992).

More promising are applications that take advantage of MC moves along with molecular dynamics (MD) simulation steps. The simplest such application uses a very short MD trajectory as a basis for a MC move. In this case, the timestep for the MD is larger than typically used in longer MD trajectories. This inserts the possibility of a larger scale conformational change, that is accepted or rejected by a metropolis criterion. Generally though MD simulation, as described earlier in this chapter, effectively samples the "fast" degrees of freedom, such as local torsional transitions, but does not, in reasonable computational time, sample large scale conformational changes or long range cooperative changes in systems. However, using modern MC methods it is possible to design hybrid MC–MD simulations that can potentially improve sampling of configurations of lipids in bilayers. Three such examples, Configurational Bias Monte Carlo (CBMC), Replica Exchange Monte Carlo (REMC), and Gibbs Ensemble Monte Carlo (GEMC) are described below.

Combined Configurational Bias Monte Carlo – Molecular Dynamics Simulation

It was shown by Chiu et al. in 1999 (1999a) that bilayers equilibrate far more efficiently if standard MD is interspersed with CBMC steps. CBMC steps are designed to move the system to a different region of phase space, guided by thermodynamic equilibrium sampling bias. More importantly they have the potential to "jump" molecules over energy barriers that may not be overcome during a long continuous MD run. There are a number of "smart" MC methods which have been applied to similar problems in liquid physics (Allen and Tildesley 1989). These methods either sample preferentially (for example, move molecules near a perturbant more often) or bias the MC moves according to the intermolecular force or some other quantity such as the virial. For a system of flexible chain molecules force-bias MC algorithms would be difficult to apply because the net force will not be constant over the entire chain, and resulting displacements would either violate bonding constraints or would only be optimal for a portion of the chain. The Configurational Bias Monte Carlo method (CBMC), due to Siepmann and Frenkel (2002), attempts to choose MC steps for entire chains or molecules which have a higher likelihood of producing energetically favorable conformations. The procedure has as its basis the early work of Rosenbluth and Rosenbluth (1955) on MC simulation of a long polymer chain on a cubic lattice.

Each CBMC step in a lipid simulation consists of the following procedure (Chiu et al. 1999a):

- Pick a single lipid at random and pick one of the two hydrocarbon chains (in this application head groups and waters are not moved in the CBMC). Pick a bond on the chosen chain at random.

- Generate a large number (Chiu et al. 1999a) of trial positions for the atom below the chosen bond (or above, in the case of head groups). Calculate the configuration energy for each trial, and the weight $w(n) = \sum_j \exp(-\beta E_j)$, where β is the inverse of Boltzmann's constant times the absolute temperature, and E_j is the energy of the atom at the jth trial position.

- Pick one of the trial positions k with probability $\exp(-\beta E_k)/w(n)$.

- Repeat the above process until the end of the chain is reached. Calculate the "Rosenbluth Weight" $W(n) = \prod_{atoms} w(n)$.

- Calculate the "old Rosenbluth Weight", $W(o)$, of the initial configuration by repeating the above procedure with one of the trials at each atom position being the original position of that atom.

- Accept the new configuration of the chain with probability
$$p = \min\left[W(n)/W(o), 1\right]$$

- If configuration is rejected, reset coordinates of chosen chain or head group to original values.

Seipmann and Frenkel (2002) and other texts describe the proof that the above procedure satisfies the requirements that configuration averages over states converge towards equilibrium thermodynamic averages. The CBMC procedure is carried out at the fixed box dimensions from a previous MD trajectory at the same temperature as the MD run.

The end result of the CBMC procedure is (if accepted) a move to an entirely new chain configuration which fits well (in a MC sense) with the local environment of that chain. In a traditional MC procedure, in which only one or two C–C bonds are rotated to a new dihedral position, such an optimal *whole-chain* conformation would be generated only very rarely. Seipmann and Frenkel (2002) have tested the CBMC algorithm against classical Monte Carlo and other techniques for systems of alkane chains of different lengths. In all cases the CBMC method provided faster convergence and better accuracy, in some cases by many orders of magnitude.

By itself CBMC will not efficiently simulate a lipid bilayer because water and polar groups need to be included. Chiu et al. (1999a) used the combined CBMC–MD protocol described above to fully sample lipid molecular conformations. This procedure exploited complementary features of MD and CBMC. CBMC moves across the energy barriers associated with torsion angles in the phospholipid structure more readily and efficiently than MD or conventional MC. Efficiency is achieved via reduced degrees of freedom, focusing on important transitions in configuration space, in particular torsion angle transitions between relative energy minima in torsion angle phase space. In this way CBMC is enabled to explore large regions of configuration space in a coarse–grained fashion, with convergence to a Boltzmann distribution. On the other hand MD explores local regions of configuration space more thoroughly than CBMC because it replicates the continuous motions of the system. In the latest version of our method all translational motions (which are less capable of being expressed in reduced degrees of freedom than the torsion angle transitions) are produced by the MD. MD is guaranteed to ultimately converge to a Maxwell–Boltzmann distribution for all degrees of freedom if either: (i) the simulation is started near equilibration, or (ii) the simulation is carried out for a very long time. When these procedures are alternated the CBMC calculation permits the system to make substantial conformational changes provided the results of those changes are thermodynamically acceptable, while the subsequent MD calculation explores the region of conformation space near the region that CBMC has moved and drives all translational moves. In the combined simulations this procedure is analogous to a series of small thermodynamic changes in the system, moving it through phase space towards a thermodynamic equilibrium state. Further interspersing longer continuous MD production runs with CBMC steps allow for a more thorough configuration space sampling. This procedure will be particularly effective in heterogeneous systems where lipid chains must interact with more rigid molecular interfaces as presented by cholesterol or membrane proteins.

Figure 1.22 shows the flowchart for a combined MD–CBMC equilibration procedure from Chiu et al. (1999a). Figure 1.23 shows a plot of the time

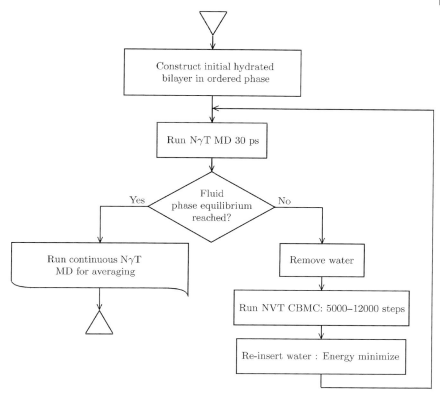

Fig. 1.22 Flowchart for combined MD–CBMC simulation procedure for equilibration of a lipid simulation. In this example the MD was run in a combination of NVT and NγT ensembles, but the procedure is applicable if NPT is used instead of NγT, although the short NVT simulations that follow each CBMC phase should be retained. Reprinted from Chiu et al. (1999a).

evolution of the total potential energy of an MD–CBMC equilibration run for a pure DPPC bilayer (Chiu et al. 1999a). The initial state for this run was an ordered state similar to that shown in Fig. 1.7. The figure shows an $\approx 30\%$ speedup in the evolution from to a state where the energy becomes constant, within fluctuations, over the remaining simulation time for the NID-CBMC simulation, compared to the MD-only simulation.

The entire combined equilibration process may be viewed as a series of "quasistatic" steps in the thermodynamic sense. As in a textbook expansion of a system, the process is carried out in a large number of slow, small steps so that equilibrium is never disturbed from the beginning to the end of the expansion. In experiments, such as monolayer expansions, one ideally expands or compresses the system by moving the monolayer barrier by a small amount and then stopping barrier motion to let the system equilibrate (or

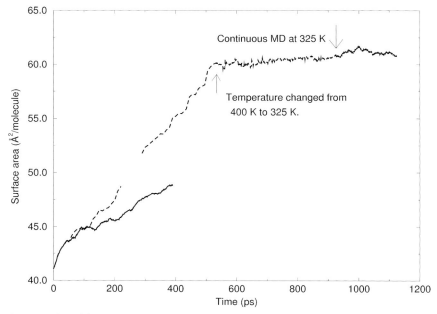

Fig. 1.23 Plot of the time evolution of the total surface area of a DPPC lipid bilayer simulation. The simulation started in a gel phase, and evolved towards a fluid phase. The dotted curve represents a combined MD–CBMC simulation while the solid curve represents an MD simulation. Reprinted from Chiu et al. (1999a).

simply move the barrier very slowly, which is a limiting case of the stop-start procedure). In our case the system moves from its initial volume towards a final fluid-phase volume in a series of MD steps. On the other hand the majority of torsion angle transitions, and hence the largest changes in the hydrocarbon chain order parameters, take place during the CBMC steps. In this procedure, as we have implemented it, the CBMC is run in an NVT ensemble using the fixed box size and temperature from the end of the prior MD step. Both the MD and the CBMC move the system reversibly through its phase space.

Replica Exchange

A different computational approach of interest for heterogeneous bilayer simulations is Parallel Tempering (PT) (Frenkel and Smit 2002) alternatively called Replica Exchange Molecular Dynamics (REMD) when applied to MD simulations (Sugita and Okamoto 1999). In a REMD or PT simulation several identical copies of the system are simulated, each under slightly different thermodyamic conditions. Probably the most common parameter which is varied is the temperature. If the system has a complex energy landscape with many local minima, simulations run at higher temperature are more likely to "hop" from one minimum to another in configuration space. In PT/REMD one periodically stops the simulations and attempts an exchange of a randomly selected pair of the systems. The exchange is accepted or rejected according to a Metropolis-like criterion. The Monte Carlo-based PT method has been applied to zeolites (Falconi and Deem 1999) and to a Lennard–Jones fluid phase equilibrium study (Yan and de Pablo 2000). REMD has been applied to protein structure prediction and folding by a number of groups including Sugita and Okamoto (1999), Garcia et al. (2002) and Rhee et al. (2004). An excellent and convincing example of PT applied to a particle in a one-dimensional multiple-minimum potential is given in the Frenkel and Smit monograph (2002). PT/REMD could be run by using separate subsets of available linux clusters for the various simulations and making the system exchange moves "by hand", or they could be set up to run in a grid environment, with separate grid nodes running the individual simulations and then communicating to perform the system exchange trial moves. The major drawback of REMD is the requirement for a large number of replicas (the energy distributions of neighboring replicas must have sufficient overlap or all replica exchange moves will be rejected), and therefore a very large number of processors. This approach is becoming increasingly important as more researchers gain access to large "farms" of linux clusters. Then a GRID environment could be used to run the multiple simulations on separate sets of nodes possibly at remote sites.

Semi-Gibbs Ensemble MC

In order to simulate systems with multiple phases and/or types of molecules, it is necessary to attempt moves that transfer molecules between phases. Within standard molecular dynamics or Monte Carlo, this type of transfer will not occur unless prohibitively long simulations are run. One alternative approach is based on the Semigrand Gibbs Ensemble Monte Carlo (SGEMC) method. The SGEMC procedure has been applied to the simulation of phase equilibria in simple systems and mixtures (Kofke and Glandt 1988). In the SGEMC ensemble the temperature, pressure and differences in chemical potentials of the molecules are fixed. In a SGEMC step a randomly chosen molecule of one type is changed into a molecule of the second species in the mixture. Moves are accepted according to

$$\mathrm{acc}(i \to j) = \min\left[1, \frac{\xi_j}{\xi_i} \exp(-\beta \Delta V(ij))\right] \qquad (1.28)$$

where ξ_i is the fugacity, $\exp(\beta\mu_i)$, of the molecule of type i with chemical potential μ_i and $\Delta V(ij)$ is the difference in potential energy of the system after the identity change move, and $\beta = 1/kT$ (Frenkel and Smit 2002). In the limit of a large number of MC moves the composition and lateral organization of the molecules evalues toward the thermodynamic equilibrium state for the given T, P, and $\Delta\mu$.

The lack of knowledge of the molecular chemical potentials is a disadvantage of the SGEMC method. However, it is possible to design MC moves that swap molecular identities between two randomly chosen lipids of different types in a mixed lipid bilayer. This process begins by randomly selecting one molecule of each type. Then morph molecule 1 into molecule 2 and vice versa. CBMC can be used to find the best fit (in a Monte Carlo sense) of the swapped molecules into the space of the original molecules. The acceptance probability will be based on the CBMC algorithm, with old weights generated from the original positions of the molecules. Since this is a double morphing move changes in chemical potential cancel out eliminating the need to know them. The high density in the interior of a lipid bilayer presents problems for any MC move that involves molecular replacement or insertion. Actual full-molecule insertion is not a viable possibility for a MC move, but the SGEMC identity-swap moves have a better chance of being accepted because the new molecule is placed into the space occupied by the old molecule. However since new and old molecules are not identical, there will always be "fitting" problems. The Configurational Bias MC chain placement procedure will help to overcome this problem by exploring the available space for the thermodynamically optimal conformation for the new molecule in the given space. After a successful swapping move traditional MD in the NPT ensemble can be run to further relax the system and explore other molecular degrees of freedom. This algorithm has been tested on a simple bilayer consisting of 200 DPPC and 200 DMPC molecules, plus 10 000 waters at a temperature of $305\,\text{K}$ (between the transition temperatures of DPPC and DMPC). Figure 1.24 shows the identity change of a single DMPC into a DPPC. This is a relatively simple change involving the use of CBMC to generate positions for the two extra methylenes needed by DPPC. The swap moves proposed herein will of course be more difficult, but we believe that this algorithm may provide a way to move molecules large distances in a simulation while retaining thermodynamic equilibrium distributions.

1.2.6
Thermodynamic Quantities, Limitations of Atomistic Simulations

In spite of continual advances in computational power the direct application of atomistic simulations to biological systems at length and time scales of bi-

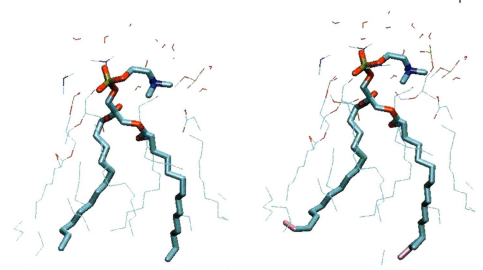

Fig. 1.24 A sample swap move of DMPC and DPPC molecules. Here the initial DMPC (left) is replaced by DPPC (right) using our new algorithm. Extra atoms added using CBMC are shown in purple.

ological significance is decades away. To illustrate the problem consider that lateral diffusion coefficients for the lipids in liquid crystalline phase, determined in experiments, are of the order of 5×10^{-12} m^2/s. This value along with the Einstein diffusion relation gives approximately a 1.7 nm root mean square (RMS) displacement for a lipid molecule in 250 ns. While 250 ns is attainable in an atomistic simulation at this time, the wall clock time for such a run is on the order of months.

Thermodynamic properties of interest such as phase changes, heat capacities, or permeabilities are also beyond the scope of atomistic simulation. Lateral phase separation or domain formation involving large numbers of molecules will not be observed in atomistic simulations any time soon. While it can be argued argued that in 250 ns it is possible to identify the initial stages of domain formation, it will be necessary to employ other modeling methods to directly observe large scale atomic rearrangements in a bilayer (Pandit et al. 2003c).

1.3
Coarse Grain Models

As a consequence of this limitation there have been a number of efforts to devise alternative "coarse grained" simulation models that retain the essential features of atomistic lipids but that have otherwise severely reduce the number of molecular degrees of freedom, thereby greatly accelerating the simulations. In this section we describe several coarse graining strategies that have the goal of extending simulations to greater scales of length and time. In each case atomistic detail is sacrificed and the choices that are made affect the properties of the coarse granied models.

1.3.1
Simulations Based on Reduced "Pseudo-Molecular" Models

One way to model lipid bilayers on larger scales of length and time is to reduce the number of degrees of freedom per molecule. This approach condenses a typical lipid molecule from about 50 (non-hydrogen) atoms to between four and twenty "pseudo-molecules" by combining atoms along the three lipid chains into larger atoms. The pseudo-molecules are constructed as chains of typically 4-10 pseudo-atoms (solvent is generally represented by single pseudo-atom molecules). After defining suitable forcefields for the coarse grained model the properties of the system are calculated using standard MD or MC. The effect of the reduced number of degrees of freedom is to allow for a time step of ≈ 5 picoseconds in MD or larger moves in MC. Thus with a coarse grained model for the lipids one can obtain an improvement of about 3–4 orders of magnitude in computational speed (Groot and Rabone 2001; Kranenburg et al. 2003a; Kranenburg et al. 2003b). Lipowski and co-workers have studied similar coarse-grained molecular models for bilayers using Monte Carlo (MC) and MD simulation (Goetz and Lipowsky 1998; Goetz et al. 1999). Marrink et al. have developed a robust forcefield for coarse grained simulations of lipid bilayers (Marrink et al. 2004) allowing for the most comprehensive simulations of lipids in this approximation to date. The model consists of mapping the non-hydrogen atoms of a typical phospholipid, DPPC, consisting of 50 non-H atoms, onto a pseudo-lipid consisting of 10 lipids. The coarse grained molecule has chains of four atoms each, a "backbone" of two atoms and a polar group of two atoms (for phosphate and choline). Water is modeled as a single sphere. All of the atoms interact with each other through 6-12 potentials, with screened coulombic forces included between head group atoms. Bond length and angle potentials are included as harmonic functions and a trigonometric potential is used for chain dihedrals. The strengths of the interactions are designed to model shielded electrostatic interactions where needed, and all interaction parameters are adjusted to fit electron density and atom distribution profiles from full atomistic simulations. A cutoff of 1.2 nm is used for all of the non-bonded interaction ranges. The reduction in degrees of freedom and the choice of parameters allowed Marrink and co-workers to run MD simulations for over 10 microseconds

(μs). This allows for the simulation of such events as pore formation (Knecht et al. 2005), vesicle fusion (Marrink and Mark 2003) and hexagonal phase formation (Marrink and Mark 2005). In an interesting application Stevens has used this type of coarse graining to show how domains form in and how they match across the two leaflets of a bilayer (Smith 2005). This level of coarse graining produces results that are of clear qualitative value in understanding the dynamical interactions of membranes on a large scale. However, there is no real connection between the atomic level interactions and the coarse grained model. The parameters for the latter were determined from experimental data rather than from atomistic forces. This approach is most useful for thermodynamic properties, but is not as useful for understanding small scale biological interactions.

1.3.2
Continuum Models

An important goal modeling of model membranes is to predict the *equilibrium* properties of the systems. Even though unrelated to much of the biological activity in living cells, equilibrium structural and thermodynamic properties of simple model lipid bilayers are essential tests of modeling work. Firstly equilibrium structural data provide insights into the interactions that control structural properties of biomembranes. Secondly equilibrium structural data provide tests that guide theoretical modeling work. Theoretical modeling work, in turn, can both yield atomistic insights into biomembrane driving forces and can aid in the interpretation of experimental data in model and biological membranes. There is now a fairly large literature on theoretical models for lipid bilayers. In this chapter we will not attempt to review this literature but will focus on current models that can be directly related to atomistic MD simulations.

Equilibrium models based on bulk properties of bilayers: A technique for the extension of atomistic simulation predictions to biological scales has been developed by Ayton et al. in a series of papers (Ayton et al. 2001; Ayton et al. 2002; Ayton and Voth 2002; Ayton et al. 2004; Ayton et al. 2006; Ayton et al. 2007). In this modeling effort the lipid bilayer is described at a mesoscopic level by a continuous elastic membrane. The elastic modulus for the membrane is calculated by either non-equilibrium MD or dissipative particle dynamics (DPD) methods, with input from the MD for the bulk properties. Ayton et al. and co-workers describe several methods by which et al. use local bulk elastic properties for their mesoscopic simulations by which they map mechanical properties from MD simulations onto coarse grained models. The coarse graining in this case is to a continuum elasticity based field.

The elastic modulus for the membrane is calculated by non-equilibrium MD or DPD methods based on input from the MD for the bulk properties. Properties of the field are then calculated by continuum simulation algorithms. When applied to lipid-cholesterol mixtures, for example, an elasticity field is coupled with a composition field. The latter is evolved in time using the Cahn–Hilliard equation (Lubensky and Chaikin 1995). Applied to a mixture of cholesterol and dimyristoylphosphatidylcholine (DMPC) this model predicts composition-curvature fluctuations that can be interpreted as microdomains rich in cholesterol and ordered DMPC. One disadvantage of this model is that the intersections are highly complex and simulation times are about the same as atomistic MD

Equilibrium models based on mean field theory: The major challenge for all coarse-graining modeling is to link the dynamical entities (pseudo-particles or fields) with accurate atomic level detailed properties of the system. A promising method for simulations which can address this problem, and that can be run at scales of micrometers and milliseconds, is Langevin Dynamics driven by free energy gradients in a locally defined molecular order (Lubensky and Chaikin 1995; Balazs et al. 2000; Peng et al. 2000). In this type of modeling a system is defined over an area or volume by a localized free energy functional of a suitably defined, localized, order parameter. Minimizing this functional over the order parameter in the absence of perturbants or external driving forces leads to the usual static Ginzburg–Landau theory for phase transitions. Dynamics are introduced via the time-dependent Ginzburg–Landau (TDGL) equation (Lubensky and Chaikin 1995). The application of this methodology to mixtures of DPPC-cholesterol is discussed below.

Equilibrium models for systems with the internal degrees of freedom and complexity of a lipid bilayer present a significant challenge. It is in principle necessary to denumerate all of the molecuar conformational states, and calculate energies of interaction between molecules for all possible states, by summing atom by atom. Clearly, some approximation scheme is needed. In 1974 Marčelja formulated a statistical mechanical model for a lipid bilayer. The model was based on the Maier Saupe model for liquid crystals (Maier and Saupe 1959). In this approach the complex inter-atomic interactions that contribute to conformation-dependent intermolecular interactions are approximated by an interaction function based on hydrocarbon chain molecular order parameters. The *molecular* order parameter is defined as the average of the segmental order parameters over the chain:

$$\psi(\boldsymbol{r}) = \sum_n s_n(\boldsymbol{r})/N \qquad (1.29)$$

where $s_n(\boldsymbol{r})$ is defined in Eq. (1.14).

The Hamiltonian function in this framework is

$$H = -\sum_{<ll'>} V_0 s_l s_{l'} \quad (1.30)$$

where the sum extends over nearest neighbor pairs of lipid chains. The basic assumption in this model is that the inter-molecular interactions can be considered to be bilinearly dependent on the average chain order parameters.

In the Mean Field approximation the Hamiltonian is written as

$$H = -\sum_{<ll'>} V_0 s_l <s_{l'}> = -<s> \sum_{<ll'>} V_0 s_l \quad (1.31)$$

where $<s>$ is the bilayer-average chain order parameter which does not depend on the molecular position in a homogeneous bilayer.

The free energy of the system is given by

$$F = -k_B T \ln Z_{tot} = -k_B T \ln \left(\sum_{all\ chains} \sum_{all\ conf} \exp\left[-\frac{H}{k_B T}\right] \right) \quad (1.32)$$

where Z_{tot} is the total partition function of the system, k_B is the Boltzmann constant, T is the absolute temperature and the summation on the right hand side is over all the possible configurations of all the lipid chains. Evaluating F in the Mean Field approximation (SCMFT) the sums in Eq. (1.32) reduce to single-molecule sums and can be evaluated explicitly. However, the presence of $<s>$ in the Hamiltonian requires that one must close the theory by solving a self-consistent equation

$$\langle <s> \rangle = \frac{\sum_{all\ conf} s_c \exp[\beta \Phi s_c]}{Z_i}$$

$$= \frac{\sum_{all\ conf} s_c \exp\left[\beta \left\{\sum_{j=1}^{\nu-c_i}(V_0\langle s_j\rangle)\right\} s_c\right]}{Z} \quad (1.33)$$

Φ is the mean field at a lipid chain due to the average interactions of all neighboring chains

$$\Phi = \sum_{j=1}^{\nu-c_i} V_o \langle s_j \rangle \quad (1.34)$$

and depends on the values of $\langle s_j \rangle$ at the neighboring sites. Equation (1.33) is a non-linear equation, which can be solved numerically. In the solution procedure in order to evaluate the partition function Z it is necessary to sum over lipid chain conformations. Marčelja (1974) did this by generating the statesusing the rotational-isomeric model on a computer. With a phenomenological choice of the coupling constant V_0 Marčelja was able to fit this theory to the temperature and enthalpy change for the main lipid bilayer chain melting phase transition. While Mean Field theory seems to be an unlikely modeling method for a system with the conformational degrees of freedom of a lipid bilayer recent applications, summarized in an earlier volume in this series by Matson and Müller (2006), show that this method is quite robust and appears to be well suited to modeling lipid bilayers, as long as one is not concerned about behavior very close to a second order critical point.

Schick and co-workers have used the Marčelja model as a base for modeling the interactions between lipids of different chain saturation states and cholesterol (Scoville-Simonds and Schick 2003; Elliott et al. 2005; Elliott et al. 2006). For the application of the model to di-unsaturated chains Scoville-Symonds and Schick (2003) reduced the effective chain length in the Marčelja model by reducing the number of CH_2 molecules that contribute to the average order in chains with double bonds by two. The single chain states that are used in the partition function evaluation are generated much the same way as was done by Marčelja except the contributions from CH_2 molecules associated with double bonds are not counted. This model also allows Scoville-Symonds and Schick to model the effect of the locations of the double bonds on the properties of the membrane. Since this model calculates equilibrium properties the phase transition temperatures can calculated as functions of the location of the CH=CH double bonds.

In the Mean Field approach based on the Marčelja model the underlying system is two-dimensional. Lipid-lipid interactions are computed as products of chain-averaged order parameters is that details of the interactions between specific chain segments on neighboring chains within the bilayer are lost. Elliot et al. (2005) developed a model that includes details of these interactions in the third, interior bilayer, dimension in order to capture details of the effect of unsaturated bonds on the bilayers. This model was again solved in the self consistent mean field approximation. Applying the model to single component DOPC and DPPC bilayers, and to mixed DOPC–DPPC bilayers of varying composition, Elliot et al. were able to calculate single lipid phase transitions and, more significantly, phase separation transitions and binary phase diagrams for the DPPC–DOPC mixtures (Elliott et al. 2006). The model was applied to DPPC–cholesterol using the same basic model and the SCMFT approach. The phase diagram they calculated was in good agreement with experiments. This set of papers demonstrates the robust applicability of SCMFT to systems as complex and non-uniform as mixed lipid bilayers.

The SCMFT models of Schick and co-workers calculates equilibrium thermodynamic properties of the lipid bilayers under consideration. In order to also extract *dynamical* behavior of mixed lipids, on a larger scale than is possible in MD simulations, SCMFT can be combined with Langevin Dynamics modeling. We describe work in this direction in the following subsection.

1.3.3
MD Based Langevin Dynamics and Mean Field Theory

A major challenge for all coarse graining modeling is to link the dynamical properties of membranes with an accurate atomic level set of intermolecular forces. For this modeling the equilibrium methods of the previous section must be extended to allow time-dependent behavior. A promising method for simulations which can address this problem, has a basis in SCMFT and can be run at scales of micrometers and milliseconds, is Langevin Dynamics driven by free energy gradients in a locally defined molecular order (Lubensky and Chaikin 1995; Balazs et al. 2000; Peng et al. 2000). In this type of modeling a system is defined over a region of appropriate dimensionality by a localized free energy functional of a suitably defined, localized, order parameter. Minimizing the free energy functional over the order parameter in the absence of perturbants or external driving forces leads to the usual static Ginzburg–Landau theory for phase transitions. Dynamics are introduced via the time-dependent Ginzburg–Landau (TDGL) equation (Lubensky and Chaikin 1995). In an earlier application of this method (Peng et al. 2000) the basic free energy functional was, based on the Landau Theory of phase transitions (Landau and Lifshitz 1980), shown in Eq. (1.35)

$$F = \int d\boldsymbol{r} \left[-a\psi(\boldsymbol{r})^2/2 + b\psi(\boldsymbol{r})^4/4 + c\nabla(\psi(\boldsymbol{r}))^2 \right] \quad (1.35)$$

where $\psi(\boldsymbol{r})$ is the order parameter for the model. This free energy functional is phenomenological and is not sensitive to the details of the system. In this subsection we describe an approach that uses coarse grained Langevin Dynamics (LD) with a free energy that is based on the Marčelja model described above. In addition this approach calculates interactions and chain conformation from libraries obtained from atomistic simulations. The method extends the predictions of atomistic simulations to 100 + microsecond time and to micrometer length scales. The use of MD data reduces the number of phenomenological parameters to one, which is used to fit the main lipid chain melting transition. The LD methodology has been applied to mixtures of dipalmitoylphosphatidylcholine (DPPC) and cholesterol (Khelashvili et al. 2005; Pandit et al. 2007a) and it can be extended to ternary mixtures using interaction functions and coupling constants calculated from the MD simulations on the approrpriate atomistic membranes.

The Langevin modeling is based on a Ginzburg–Landau approach to the study of coexisting phases (Lubensky and Chaikin 1995). One defines an order parameter which characterizes the phase transition, ψ, and a free energy functional $F(\psi, t)$. ψ can be, for example, the difference in densities of liquid and vapor phases, or the spontaneous magnetization per particle in simple systems. For a lipid bilayer an appropriate choice is the same as used by Marčelja in which ψ is the *local* average chain order parameter field:

$$\psi(\boldsymbol{r}) = \sum_n S_n(\boldsymbol{r})/N \tag{1.36}$$

where $S_n(\boldsymbol{r})$ is the C–H order parameter at carbon n for the chain at position $\boldsymbol{r}$ and N is the total number of carbons for which S_n is calculated in the chains of the lipids. The free energy then evolves in time according to the Time-Dependent Ginzburg–Landau (TDGL) equation (Lubensky and Chaikin 1995; Peng et al. 2000):

$$\partial \psi(\boldsymbol{r})/\partial t = -\Gamma \partial F(\psi(\boldsymbol{r}))/\partial \psi(\boldsymbol{r}) \tag{1.37}$$

where Γ is an "order parameter mobility". This equation treats the lipid fluid phase as a continuum. The order parameter $\psi(\boldsymbol{r})$ is *not* conserved over the bilayer (i.e. the sum of $\psi(\boldsymbol{r})$ over all lipids is not fixed). An alternative approach is to use the Cahn–Hilliard equation (Lubensky and Chaikin 1995) in which the right hand side of the above equation contains a Laplacian operator in addition to the functional derivative, conserves $\psi(\boldsymbol{r})$, but this is not appropriate in the case of fields of fluctuating chain order. If, in addition to a chain order field, the system contains discrete objects (such as cholesterol molecules) each of which has an orientation θ and a position $\boldsymbol{r}$ then θ_i and $\boldsymbol{r}_i$ evolve in time according to Langevin equations based on associated gradients in the system free energy F (Peng et al. 2000)

$$\partial \boldsymbol{r}_i/\partial t = -M \partial F/\partial \boldsymbol{r}_i + \boldsymbol{\eta}_i \tag{1.38}$$

$$\partial \theta_i/\partial t = -M' \partial F/\partial \theta_i + \zeta_i \tag{1.39}$$

where M and M' are mobilities, and η_i and ζ_i are thermal fluctuations which are derived from an appropriate Gaussian distribution.

In the adaptation of this methodology to a lipid bilayer one first defines and initializes a large (at least $\sim 300 \times 300$) two dimensional lattice network of lipid chains plus cholesterols. The lipid chain order parameters values are assigned to points on the lattice while the cholesterols, modeled as small hard rods which move continuously over the plane, are given initial random center of mass positions and orientations. The cholesterol rod length is set at 0.7 times the lattice constant to match the relative size of a cholesterol molecule relative to lipid chains. The mean field model was carried out in the following sequence:

- Define lattice size and cholesterol concentration.

- Initialize: randomly distribute chain order and rods representing cholesterols over lattice.

- Solve self consistent mean field equations for initial distribution.

- Advance time by one timestep; move cholesterols according to Eq. (1.37).

- Solve self consistent mean field equations with new cholesterol distribution.

- Go to step 4.

The purpose of this modeling is to predict the lateral organizational structure of heterogeneous membranes. To illustrate we describe how the model has been applied to DPPC–cholesterol bilayers over a range of cholesterol concentrations and temperatures (Pandit et al. 2007a). A key to this work is that *the interaction energies between DPPC and cholesterol were calculated from MD simulation data*. Predictions of the model are therefore extensions of MD simulation predictions, within the approximations that underly the mean field model itself. The mobilities used in Eqs. (1.37)–(1.39) are related to the experimental diffusion constant D by: $M_r = D/k_b T$, $M_\theta \approx (10\text{--}100) \times M_r$, and $\Gamma \approx 10 M_r$. The dimensionless simulation timestep $\Delta\tau$ is related to the real-time step length Δt by $\Delta\tau = \frac{D}{a^2}\Delta t \approx 10^7 \Delta t$, where a is the lattice constant, set in preliminary work at 0.65 nm. This allows for nanosecond or greater real time step sizes with reasonable $\Delta\tau$ for the coarse grained simulation. For higher cholesterol concentration smaller timesteps are needed but the self-consistent equations converge faster in this case, compensating for the shorter timestep. Figure 1.25 shows the results of the application of this method to DPPC–cholesterol mixtures.

Figure 1.25 shows color density plots of the variation of the order parameter field across the simulation cell for three different cholesterol concentrations. The density is in this case order parameter density, with darker colors representing higher chain order. The small lines represent the locations and orientations of cholesterols and are inevitably surrounded by regions of high relative chain order. In initial investigation of the SCMFT modeling approach, we considered DPPC-cholesterol mixtures (Khelashvili et al. 2005). Based on MD simulations we derived the SCMFT model for DPPC–cholesterol system described in Section 1.3.3. Simulations for DPPC+cholesterol systems at 5, 10, 15, 20 ,25, 30 and 35% cholesterol concentration and on a wide range of temperatures from 303K to 328K were performed. After 20 μs of SCMFT simulation at each temperature and cholesterol concentration, predictions of this model were compared to experimental data as described below.

In Fig. 1.26(a) we show a plot of the average order parameter as a function of temperature for various chol concentrations within this model. In the absence of chol the model clearly exhibits a first order phase transition at 315 K. The order parameters at temperatures below and above phase tran-

68 | *1 Simulations and Models of Lipid Bilayers*

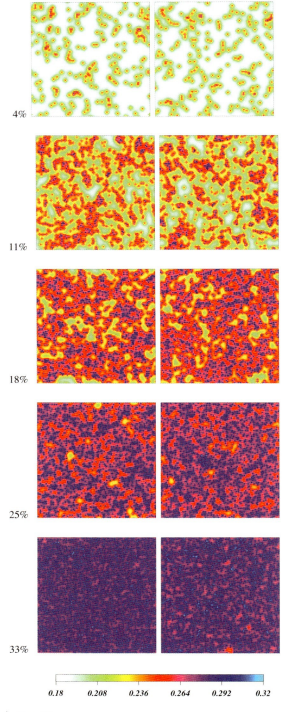

Fig. 1.25 Legend see p. 69

sition are similar to the values expected in gel and liquid crystalline phases of DPPC respectively. A closer examination of Fig. 1.26(a) reveals that with increasing cholesterol concentration the phase transition diminishes and completely vanishes for concentrations above 15%. At concentrations above 25% the system has uniform order across all the temperatures investigated here, representing the liquid ordered (β) phase. Similar results are reported by the NMR and DSC experiments of Huang et al. (1993). Figures 1.26(b) to (e) show a comparison of average order parameters obtained using our model, and the NMR quadrupolar shift reported by Huang et al. In their study Huang et al. report the shift for sixth carbon atoms in DPPC hydrocarbon chains. Since the order parameter of the sixth atom is on the plateau region of the order parameter profile, one expects the average order parameter over the entire chain to be lower than the order parameter of the sixth atom. Figures 1.26(b–e) show distributions of order parameters calculated from the model and the calculated average order parameters show remarkable agreement with the NMR data for various temperatures and cholesterol concentrations.

In order to quantify the existence of regions within the model field of different levels of chain order one can obtain the distribution of chain over the order parameter field order by a binning procedure. For certain temperature and cholesterol concentrations coexisting regions of different order are found, as revealed by a bimodality in the distribution. This property can be used to identify different regions of a "phase diagram". Figure 1.27 shows distributions of order parameters at 5%, 25% and 30% of chol concentration at various temperatures. The distributions show peaks around the order parameter values corresponding to gel and liquid crystalline order. For certain values of concentrations and temperatures these distributions show two peaks indicating coexistence of two "phases" for example, at 50 C and 25% cholesterol concentration the system clearly has multiple peaks in order parameter distribution.

By examining all distributions at various temperatures and cholesterol concentrations a plot can be drawn that outlines the regions in which the distribution of order parameters is bimodal as revealed in Fig. 1.28. This figure is fully consistent with the phase diagram proposed by Vist and Davis (1990).

Fig. 1.25 Order parameter density plots of the average local DPPC chain order parameters for two different cholesterol concentrations, at times $t = 0$ (left), and $t = 20\mu s$ (right). Color scale indicates magnitude of local order parameters on the lattice. Analysis of the plots indicates that cholesterol creates local ordered regions which merge into larger regions of higher order as the cholesterol concentration increases. At concentrations between 12 and 20% we observe dynamical coexistence of regions of high order (blue-red) and regions of relatively lower order (pale green). Reprinted from Pandit et al. (2007).

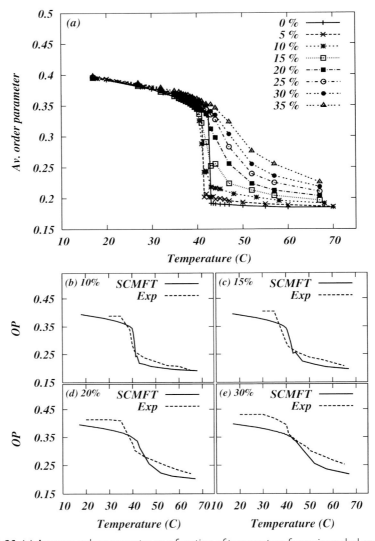

Fig. 1.26 (a) Average order parameter as a function of temperature for various chol concentrations. Comparison of the average order parameter in the model with the order parameter of the sixth carbon atom reported by Huang et al. for (b) 10% chol, (c) 15% chol, (d) 20% chol, and (e) 30% chol. Reprinted from Pandit et al. (2007c).

In particular, both model and experiment show non-overlapping regions having different degrees of chain order at the same cholesterol concentrations and temperatures. At temperatures below 315 K, coexisting regions of gel and intermediate order ($gel + \beta$ region). At temperatures above 315 K, there are non-overlapping regions of intermediate and low levels of order ($L_\alpha + \beta$

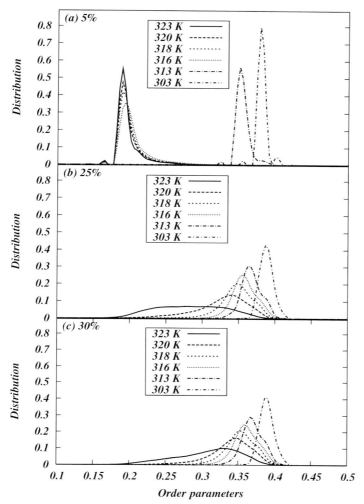

Fig. 1.27 Distribution of order parameter values as a function of temperature, at three cholesterol concentrations. Bi-modality in a plot reveals regions within the bilayer of different degrees of lipid order at that temperature and cholesterol concentration. (a) 5% cholesterol, (b) 25% cholesterol, (c) 3% cholesterol. Bi-modality is seen at lower cholesterol concentration over a range of temperatures that decreases as the cholesterol concentration increases.

region). The agreement between the model diagram and the Vist–Davis diagram is shown in Fig. 1.28, and is significant in that no parameters in the model were set phenomenologically to fit the experimental plot.

The model was tested further against experiment by computing the specific heat as function of temperature. The specific heat at constant pressure was calculated by

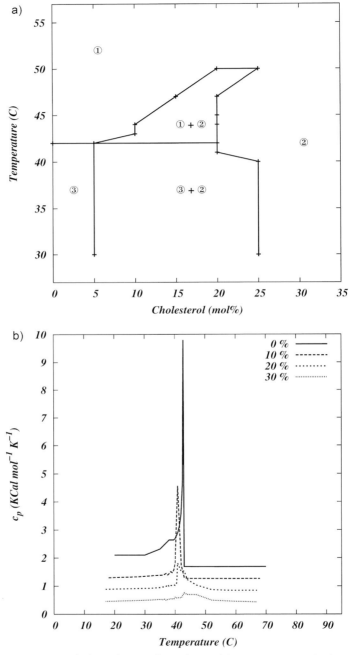

Fig. 1.28 (a) Computed phase diagram from the model. Dots represent the locations of peaks in distributions of chain order in Fig. 1.27. Lines are guides to the eye. (b) Plots of specific heat as a function of temperature for several cholesterol concentrations. Reprinted with permission from Pandit et al. (2007c).

$$c_p = T\left(\frac{\partial S}{\partial T}\right)_p$$

$$= \sum_i \left[\frac{\beta}{T}\frac{\partial <s_i>}{\partial \beta}\left(\sum_\alpha V_{lc}(1-\Delta\sin\theta_{\alpha,i}) + V_0\sum_{j=1}^{\nu_i} s_j\right)\right.$$

$$\left. -\frac{1}{T}\sum_i \Phi_i <s_i>\right] \quad (1.40)$$

where the $\partial <s_i>/\partial\beta$ in the expression was obtained by solving the self-consistent equation

$$\frac{\partial <s_k>}{\partial \beta} = \left[\Phi_k + \beta\sum_{j=1}^{\nu_k} V_0 \frac{\partial <s_j>}{\partial \beta}\right]\Delta s_k^2 \quad (1.41)$$

with $\Delta s_k^2 = <s_k^2> - <s_k>^2$. Figure 1.28 shows the specific heat as a function of temperature for the systems with 0%, 10%, 20%, and 30% concentration of cholesterol. A comparison with Fig. 4 of Huang et al. (1993) shows that the model DSC curves are in good agreement with the experimental DSC curves. One important prediction of this model is that there is no first order phase separation transition in this model. All changes in order are continuous, so that the bimodal distributions of order do not represent coexisting equilibrium phases. This conclusion is independently supported by recent re-analysis of NMR data by McConnell and Radhakrishnan (2006).

The SCMFT approach is readily generalizable to ternary and higher mixtures of lipids. In order to model the mixed lipid bilayer a new field that describes the local molecular composition of the bilayer is introduced on the lattice at each site r, for example for sphingomyelin (SM) and dioleylphosphatidylcholine (DOPC) one has:

$$\Phi(r) = \phi_{SM}(r) - \phi_{DOPC}(r) \quad (1.42)$$

where ϕ_x represents the concentration of species x, at the point r and Φ is then the local difference in concentration. One can construct a free energy functional which includes the interactions of all of the lipids in the mixture, all using maximum input from
 simulations. In addition to propagating cholesterols through the model plane we now also propagate SM and DOPC, in this case using a Cahn–Hilliard equation (in order to conserve the total number of atoms, this equation must be used instead of the Langevin equation) (Lubensky and Chaikin 1995). The functional contains coupling terms between different molecules which will be calculated from simulations and will be used to predict the lateral organization of membranes on the scale of the TDGL method. Figure 1.29 shows calculated ordering in mixtures of SM-DOPC-cholesterol ob-

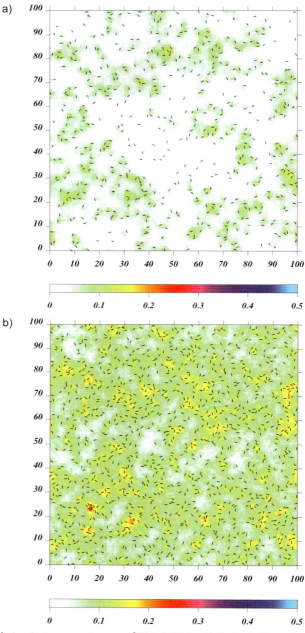

Fig. 1.29 Ordering in ternary mixtures of SM, DOPC and cholesterol. (a) 1:1 mixture of DOPC and SM with 10% cholesterol. (b) 1:1 mixture of DOPC and SM with 20% cholesterol. Darker color indicates higher order, and therefore a greater concentration of SM, while small segments represent cholesterol molecules.

tained using a preliminary estimate of the salient coupling constants from MD simulations and after an LD-SCMFT run of 100 μs.

It is evident that cholesterol induces the formation of regions of increased order in the field. In this model increased order is correlated with an enhanced concentration of sphingomyelin. By a quantitative examination of the distribution of order over the field, it is possible to map out a ternary phase diagram for this mixture. This work is in progress.

1.4 Summary

In this chapter we have described modeling of lipid bilayers. We started at the atomistic level, where molecular dynamics or Monte Carlo are the methods capable of providing details of atomistic phenomena at this level. Molecular dynamics, while simple to describe conceptually, must be applied with caution to successfully model a hydrated lipid bilayer. The complex nature of the interatomic interactions means that individual atoms will experience large forces from chemical bonds, dihedral forces, steric forces, electrostatic forces and weak van der Waals attractions (which nevertheless sum to an appreciable contribution). While the forces are large, their interplay leads to cancellations in many cases. One excellent example is the dihedral angle distribution in a chain. Consideration of the dihedral potential in isolation leads to a model (the rotational isomeric model) in which chains take on mainly conformations with dihedrals at one of the three minima of this potential. However, simulations show that the presence of 6–12 forces allows dihedral angles to take on values that are displaced from the dihedral potential wells. Further, small differences in each dihedral along a chain will propagate to large differences in chain conformations.

The development of MD forcefields for lipid bilayers is a process that cannot be overlooked. Depending on whether hydrogen atoms are to be included or excluded the forcefields will be quite different. In general, both types of forcefields are of the highest quality when successfully tested against experimental smaller "model compounds" in independent simulations before being applied to a lipid simulation. In this vein the model for molecular water is an important consideration. Most current simulations employ water models that consist of fixed point charges for the oxygen and the hydrogens, and a van der Waals force centered at the oxygen. However, new developments and faster computers may allow for the use of greatly improved water models in the near future.

We discussed statistical mechanical ensembles and their implementation in MD simulations. A lipid bilayer MD simulation should be performed (usually) in an NPT ensemble, so that pressure and temperature coupling algorithms must be applied. The size of the simulated bilayer is another important consideration. While smaller simulations can of course be run far more quickly, they may contain artifactual properties. For simulation cells consisting of 60 or fewer lipids in each leaflet most of the lipids will be at the perimeters of the cell, where periodic boundary artifacts enter in. For small simulations it is also necessary to cut off intermolecular interactions at shorter distances and Ewald summation is done over a smaller periodic system.

The process of constructing an initial bilayer to begin simulations, and the equilibration of that bilayer, are described in this chapter. We point out several ways by which an initial bilayer may be constructed and we describe the generally used equilibration procedures. One unique addition that improves equilibration is the use of Monte Carlo steps interspersed with MD steps during the initial phases of the simulation. After equilibration a production MD run is carried out for up to 100 ns, or more, generating a continuous phase space trajectory for the bilayer and the surrounding water. From the trajectory one calculates properties of the simulated bilayer that can be tested against experiment and that offer predicted new modes of behavior. Many of the properties of interest are described in this chapter.

While atomistic simulations are of great interest and provide many new insights they are limited by the relatively slow rate of diffusion of lipids in a bilayer to time scales where individual molecules do not move very far laterally. However, atomistic simulations do provide clues for larger scale thermodynamic properties of bilayers in the local interactions. We describe efforts to use input from MD simulations to construct models based on self consistent mean field theory that can describe larger scale lateral organizational development over time and across the membrane. Self Consistent mean field theory, based on models with parameters calculated from atomistic simulations has, as we illustrated in this chapter, great promise for understanding the properties of lipid bilayers of complex composition. The long range goal of this type of modeling is to begin to understand properties of membranes of biological composition.

In conclusion, this chapter provides an overview of many aspects of the modeling of lipid membranes. The goal is of course to better understand biological phenomena in membranes, and progress is being made in that direction. But also there are new physical insights that are emerging from the modeling studies of this type of soft matter. These insights may at some point be of use in other areas of soft matter science.

References

Allen, M. P. and Tildesley, D. J., 1989, *Computer Simulation of Liquids*. Oxford University Press. Oxford

Ayton, G., Bardenhagen, S., Sulsky, D., and Voth, G. A., 2001, *J. Chem. Phys.* **114**, 6913.

Ayton, G., Smondyrev, A. M., Bardenhagen, S., McMurtry, P., and Voth, G. A., 2002, *Biophys. J.* **82**, 1026.

Ayton, G. and Voth, G. A., 2002, *Biophys. J.* **83**, 3357.

Ayton, G. S., MacWhirter, J. L., and Voth, G. A., 2004, *Biophys. J.* **87**(50), 3242.

Ayton, G. S., MacWhirter, J. L., and Voth, G. A., 2006, *J. Chem. Phys.* **124**(50), 64906.

Ayton, G. S., Noid, W. G., and Voth, G. A., 2007, *Curr. Op. Struct. Biol.* **17**(50), 192.

Balazs, A., Ginsburg, V. V., Qiu, F., Peng, G., and Jasnow, D., 2000, *J. Phys. Chem. B* **104**, 3411.

Baxter, R. J., 1982, *Exactly Solved Models in Statistical Mechanics*. Academic Press.

Berendsen, H. J. C., van der Spoel, D., and van Drunen, R., 1995, *Comp. Phys. Comm.* **91**, 43.

Berger, O., Edholm, O., and Jähnig, F., 1997, *Biophys. J.* **72**, 2002.

Bhide, S. Y., Zhang, Z., and Berkowitz, M. L., 2007, *Biophys. J.* **92**, 1284.

Binder, K., 1986, *Monte Carlo Methods in Statistical Physics*. Springer-Verlag.

Bodin, S., Soulet, C., Tronchere, H., Sie, P., Gachet, C., Plantavid, M. P., and Payrastre, B., 2005, *J. Cell Sci.* **118**, 749.

Bukowski, R., Krzysztof, S., Groenenboom, G., and van der Avoird, A., 2007, *Science* **315**(14), 1249.

Carlson, J. M. and Sethna, J. P., 1987, *Phys. Rev. A* **36**, 3359.

Cascales, J. J. L., Smith, B. D., Gonzales, C., and Marquez, M., 2006, *J. Phys. Chem. B* **110**, 2358.

Cevc, G., 1990, *Biochim. Biophys. Acta* **1031**(3), 311.

Chiu, S.-W., Clark, M., Subramaniam, S., Scott, H. L., and Jakobsson, E., 1995, *Biophys. J.* **69**, 1230.

Chiu, S.-W., Clark, M. M., Jakobsson, E., Subramaniam, S., and Scott, H. L., 1999a, *J. Comp. Chem.* **20**, 1153.

Chiu, S.-W., Clark, M. M., Jakobsson, E., Subramaniam, S., and Scott, H. L., 1999b, *J. Phys. Chem. B* **103**, 6323.

Chiu, S.-W., Jakobsson, E., Mashl, R. J., and Scott, H. L., 2002, *Biophys. J.* **83**(4), 1842.

Chiu, S.-W., Jakobsson, E., and Scott, H. L., 2001a, *J. Chem. Phys.* **114**, 5435.

Chiu, S.-W., Jakobsson, E., and Scott, H. L., 2001b, *Biophys. J.* **80**(3), 1104.

Chiu, S.-W., Vasudevan, S., Jakobsson, E., Mashl, R. J., and Scott, H. L., 2003, *Biophys. J.* **85**(6), 3624.

Christy, W. W., 2003, *Lipid Analysis 3rd Edition*. Oily Press, Bridgewater UK.

Coe, T. J. and Scott, H. L., 1992, *Biophys. J.* **42**, 219.

Cramer, C. J., 2006, *Essentials of Computational Chemistry: Theories and Models*. John Wiley & Sons, West Sussex, England.

Damodoran, K. V. and Merz, K., 1994, *Biophys. J.* **66**, 1076.

de Vries, A. H., Mark, A. E., and Marrink, S.-J., 2004, *J. Am. Chem. Soc.* **126**(14), 4488.

de Vries, A. H., Yefimov, S., Mark, A. E., and Marrink, S.-J., 2005, *Proc. Natl. Acad. Sci. USA* **102**, 5392.

Dolan, E., Venable, R., Pastor, R. W., and Brooks, B., 2002, *Biophys. J.* **73**(23), 2317.

Doniach, S., 1979, *J. Chem. Phys.* **70**, 4587.

Douliez, J.-P., Léonard, A., and Dufourc, E. J., 1995, *Biophys. J.* **68**, 1727.

Edholm, O. and Nagle, J. F., 2005, *Biophys. J.* **89**(3), 1827.

Eisenberg, M., Gresalfi, T., Riccio, T., and McLaughlin, S., 1979, *Biochem.* **18**(23), 5213.

Elliott, R., Katsov, K., Szleifer, I., and Schick, M., 2005, *J. Chem. Phys.* **122**, 044904.

Elliott, R., Szleifer, I., and Schick, M., 2006, *Phys. Rev. Lett.* **96**, 098101.

Essmann, U., Perera, L., Berkowitz, M. L., Darden, T., Lee, H., and Pedersen, L. G., 1995, *J. Chem. Phys.* **103**(19), 8577.

Falconi, M. and Deem, M., 1999, *J. Chem. Phys.* **110**, 1754.

Feller, S. E. and Pastor, R. W., 1996, *Biophys. J.* **71**(23), 1350.

Filippov, A., Orädd, G., and Lindblom, G., 2003, *Biophys. J.* **84**(5), 3079.

Frenkel, D. and Smit, B., 2002, *Understanding Molecular Simulation From Algorithm to Applications*, Vol. 1 of *Computational Science Series*. Academic Press.

Frisch, M. J., Trucks, G. W., Schlegel, H. B., Scuseria, G. E., Robb, M. A., Cheeseman, J. R., Montgomery, Jr., J. A., Vreven, T., Kudin, K. N., Burant, J. C., Millam, J. M., Iyengar, S. S., Tomasi, J., Barone, V., Mennucci, B., Cossi, M., Scalmani, G., Rega, N., Petersson, G. A., Nakatsuji, H., Hada, M., Ehara, M., Toyota, K., Fukuda, R., Hasegawa, J., Ishida, M., Nakajima, T., Honda, Y., Kitao, O., Nakai, H., Klene, M., Li, X., Knox, J. E., Hratchian, H. P., Cross, J. B., Bakken, V., Adamo, C., Jaramillo, J., Gomperts, R., Stratmann, R. E., Yazyev, O., Austin, A. J., Cammi, R., Pomelli, C., Ochterski, J. W., Ayala, P. Y., Morokuma, K., Voth, G. A., Salvador, P., Dannenberg, J. J., Zakrzewski, V. G., Dapprich, S., Daniels, A. D., Strain, M. C., Farkas, O., Malick, D. K., Rabuck, A. D., Raghavachari, K., Foresman, J. B., Ortiz, J. V., Cui, Q., Baboul, A. G., Clifford, S., Cioslowski, J., Stefanov, B. B., Liu, G., Liashenko, A., Piskorz, P., Komaromi, I., Martin,

R. L., Fox, D. J., Keith, T., Al-Laham, M. A., Peng, C. Y., Nanayakkara, A., Challacombe, M., Gill, P. M. W., Johnson, B., Chen, W., Wong, M. W., Gonzalez, C., and Pople, J. A., 2004, *Gaussian 03, Revision C.02*, Gaussian, Inc., Wallingford, CT, 2004.

Goetz, R., Gompper, G., and Lipowsky, R., 1999, *Phys. Rev. Lett.* **82**, 221.

Goetz, R. and Lipowsky, R., 1998, *J. Chem. Phys.* **108**, 7397.

Gompper, G. and Schick, M., 2006, *Soft Matter, Vol. 1: Polymer Melts and Mixtures*. Wiley-VCH.

Greenwood, A. I., Tristram-Nagle, S., and Nagle, J. F., 2006, *Chemistry and Physics of Lipids* **143**, 1.

Groot, R. D. and Rabone, K. L., 2001, *Biophys. J.* **81**, 725.

Hofsäß, C., Lindahl, E., and Edholm, O., 2003, *Biophys. J.* **84**(4), 2192.

Huang, T.-H., Less, C. W. B., Das Gupta, S. K., Blume, A., and Griffin, R. G., 1993, *Biochem.* **32**(48), 13277.

Jorgensen, W., Chandrasekhar, J., Madura, J., Ampey, R., and Klein, M., 1983, *J. Chem. Phys.* **79**(14), 926.

Khelashvili, G. A., Pandit, S. A., and Scott, H. L., 2005, *J. Chem. Phys.* **123**, 034910.

Klauda, J. B., Kucerka, N., Brooks, B. R., Pastor, R. W., and Nagle, J. F., 2006, *Biophys. J.* **90**(8), 2796.

Knecht, V., Müller, M., Bonn, M., Marrink, S.-J., and de Vries, A. H., 2005, *J. Chem. Phys.* **122**, 0224704.

Kofke, D. A. and Glandt, E. D., 1988, *Mol. Phys.* **64**, 1105.

Kranenburg, M., Ventrouli, M., and Smit, B., 2003a, *Phys. Rev. E* **67**, 60901.

Kranenburg, M., Ventrouli, M., and Smit, B., 2003b, *J. Phys. Chem. B* **107**, 11491.

Kucerka, N., Tristram-Nagle, S., and Nagle, J. F., 2005, *J. Membr. Biol.* **208**, 193.

Kucerka, N., Tristram-Nagle, S., and Nagle, J. F., 2006, *Biophys. J.* **90**, L83.

Kuwajima, S., Noma, H., and Akasaka, T. (eds.), 1994, *Proc. 4th Symp. Comp. Chem.*, Tokyo, Japan. Japan Chemistry Program Exchange.

Landau, L. D. and Lifshitz, E. M., 1980, *Statistical Physics 3rd Edition Part 1; Course of Theoretical Physics, Vol. 5*. Pergamon Press, 2nd rev. edn.

Lubensky, T. and Chaikin, P., 1995, *Principles of Condensed Matter Physics*. Cambridge University Press.

Macdonald, P. M. and Seelig, J., 1988, *Biochem.* **27**(18), 6769.

MacKerell Jr, A. D., 2004, *J. Comp. Chem.* **25**(12), 1505.

Maier, W. and Saupe, A., 1959, *Z. Naturforsch. A* **14**, 882.

Makino, K., Yamada, T., Kimura, M., Oka, T., Ohshima, H., and Kondo, T., 1991, *Biophys. Chem.* **41**, 175.

Marrink, S.-J., de Vries, A. H., and Mark, A. E., 2004, *J. Phys. Chem. B* **108**, 750.

Marrink, S.-J. and Mark, A. E., 2003, *J. Am. Chem. Soc.* **128**, 11144.
Marrink, S.-J. and Mark, A. E., 2005, *Biophys. J.* **87**, 3894.
Marčelja, S., 1974, *Biochim. Biophys. Acta* **367**, 156.
Marx, D. and Hutter, J., 2000, in *Modern Methods and Algorithms of Quantum Chemistry, Proceedings, Second Edition, John von Neumann Institute for Computing, NIC Series, Vol. 3*, pp. 329 – 477, Jülich.
Matson, M. W., in: Gompper, G. and Schick, M., 2006, *Soft Matter, Vol. 1: Polymer Melts and Mixtures*. Wiley-VCH.
McConnell, H. and Radhakrishnan, A., 2006, *Proc. Natl. Acad. Sci. USA* **103**(5), 1184.
McCullough, S. and Scott, H. L., 1990, *Phys. Rev. Lett.* **65**, 931.
McLaughlin, S., 1989, *Annu. Rev. Biophys. Biophys. Chem.* **18**, 113.
Merz Jr., K. M. and Roux, B. (eds.), 1996, *Biological Membranes: A Molecular Perspective from Computation and Experiment*. Birkhauser.
Metropolis, N., Rosenbluth, A. W., Rosenbluth, M. N., Teller, A. N., and Teller, E., 1953, *J. Chem. Phys.* **21**(12), 1087.
Mukhopadhyay, P., Monticelli, L., and Tieleman, D. P., 2004, *Biophys. J.* **86**, 1601.
Müller, M., in: Gompper, G. and Schick, M., 2006, *Soft Matter, Vol. 1: Polymer Melts and Mixtures*. Wiley-VCH.
Myrzin, K., Rog, T., and Pasenkiewicz-Gierula, M., 2005, *Biophys. J.* **88**, 1090.
Nagle, J. F., 1973, *J. Chem. Phys.* **58**, 236.
Nagle, J. F. and Scott, H. L., 1978, *Biochim. Biophys. Acta* **513**, 159.
Nagle, J. F. and Tristram-Nagle, S., 2000, *Biochim. Biophys. Acta* **1469**, 159.
Niemelä, P. S., Ollila, S., Hyvönen, M., Kartunen, M., and Vattulainen, I., 2007, *PLoS Computational Biology* **3**(2), 0304.
Pandit, S., Chiu, S.-W., Jakobsson, E., Grama, A., and Scott, H. L., 2007a, *Biophys. J.* **92**, 920.
Pandit, S., Chiu, S.-W., Jakobsson, E., Grama, A., and Scott, H. L., 2007b, *preprint*.
Pandit, S., Khelashvili, G., Jakobsson, E., Grama, A., and Scott, H. L., 2007c, *Biophys. J.* **92**, 440.
Pandit, S. A. and Berkowitz, M. L., 2002, *Biophys. J.* **82**(4), 1818.
Pandit, S. A., Bostick, D., and Berkowitz, M. L., 2003a, *J. Chem. Phys.* **119**(4), 2199.
Pandit, S. A., Bostick, D., and Berkowitz, M. L., 2003b, *Biophys. J.* **85**(5), 3120.
Pandit, S. A., Bostick, D., and Berkowitz, M. L., 2003c, *Biophys. J.* **86**, 3743.
Pandit, S. A., Bostick, D. L., and Berkowitz, M. L., 2004a, *Biophys. J.* **86**(3), 1345.
Pandit, S. A., Jakobsson, E., and Scott, H. L., 2004b, *Biophys. J.* **87**(5), 3312.
Pandit, S. A., Vasudevan, S., Chiu, S.-W., Mashl, R. J., Jakobsson, E., and Scott, H. L., 2004c, *Biophys. J.* **87**(2), 1092.

Patra, M., Karttunen, M., Hyvönen, M., Falck, E., Lindqvist, P., and Vattulainen, I., 2003, *Biophys. J.* **84**, 3636.
Peng, G., Qiu, F., Ginzburg, V. V., Jasnow, D., and Bakazs, A., 2000, *Science* **288**, 1802.
Petrache, H. I., Feller, S. E., and Nagle, J. F., 1997, *Biophys. J.* **70**, 2237.
Reiling, S., Schlenkrich, M., and Brickmann, J., 1996, *J. Comp. Chem.* **17**, 450.
Rhee, Y. M., Sorin, E., Jayacyandran, G., Lindahl, E., and Pande, V. S., 2004, *Proc. Natl. Acad. Sci. USA* **101**, 6456.
Rinia, H. A., Snel, M. M. E., van der Eerden, J. P. J. M., and de Kruijff, B., 2001, *FEBS Lett.* **501**, 92.
Rosenbluth, M. N. and Rosenbluth, A. W., 1955, *J. Chem. Phys.* **23**, 356.
Ryckaert, J. P. and Bellemans, A., 1978, *Faraday Disc. Chem. Soc.* **66**, 95.
Sanbonmatsu, K. Y. and Garcia, A. E., 2002, *Proteins* **46**, 225.
Scott, H. L., 1975, *J. Chem. Phys.* **62**(4), 347.
Scott, H. L., 1977, *Biochim. Biophys. Acta* **469**(4), 264.
Scott, H. L., 1984, *Comments in Molecular and Cellular Biophysics* **11**, 197.
Scott, H. L., 1993, in *Cholesterol in Model Membranes*, edited by L. Finegold, pp. 197–222. CRC Press, Boca Raton.
Scott, H. L., 2002, *Curr. Op. Struct. Biol.* **12**(4), 495.
Scoville-Simonds, M. and Schick, M., 2003, *Phys. Rev. E* **67**, 098101.
Sengupta, K., Raghunanthan, V. A., and Katsaras, J., 2003, *Phys. Rev. E* **68**, 0331710.
Singer, J. J. and Nicholson, G., 1972, *Science* **175**, 720.
Smith, M. J., 2005, *J. Am. Chem. Soc.* **127**, 15530.
Stamatoff, J., Feuer, B., Guggenheim, H. J., Tellez, G., and Yamane, T., 1982, *Biophys. J.* **38**, 217.
Stillinger, F. and Rahman, A., 1974, *J. Chem. Phys.* **60**(4), 1545.
Stillinger, F. H., 1980, *Science* **209**, 451.
Sugita, Y. and Okamoto, Y., 1999, *Chem. Phys. Lett.* **314**, 141.
Sun, W.-J., Tristram-Nagle, S., Suter, R. M., and Nagle, J. F., 1996, *Proc. Natl. Acad. Sci. USA* **93**, 7008.
Tatulian, S. A., 1987, *Eur. J. Biochem.* **170**, 413.
Torbet, J. and Wilkins, M. H., 1976, *J. Theor. Biol.* **62**(12), 447.
Tristram-Nagle, S. and Nagle, J. F., 2004, *Chem. Phys. Lipids* **127**, 3.
Tu, K., Dobias, D., Blasie, J. K., and Klein, M. L., 1996, *Biophys. J.* **70**, 595.
Tu, K., Klein, M. L., and Tobias, D. J., 1998, *Biophys. J.* **75**, 2147.
Venable, R. M., Zhang, Y., Hardy, B. J., and Pastor, R. W., 1993, *Science* **262**(12), 223.
Venable, R. W., Pastor, R. W., and Brooks, B. R., 2000, *J. Chem. Phys.* **112**(23), 4822.
Vist, M. R. and Davis, J. H., 1990, *Biochem.* **29**(2), 451.

Wack, D. C. and Webb, W. W., 1988, *Phys. Rev. Lett.* **61**, 1210.
Whittington, S. G. and Chapman, D., 1966, *Trans. Faraday Soc.* **62**(12), 3319.
Wiener, M. C., King, G. I., and White, S. H., 1992, *Biophys. J.* **61**, 428.
Wiener, M. C. and White, S. H., 1991, *Biophys. J.* **59**, 162.
Wiener, M. C. and White, S. H., 1992, *Biophys. J.* **61**, 434.
Wohlert, J. and Edholm, O., 2004, *Biophys. J.* **87**(4), 2433.
Woodward, J. T. and Zasadzinski, J. A., 1997, *Biophys. J.* **72**, 964.
Xing, J. and Scott, H. L., 1992, *Biochim. Biophys. Acta* **1106**, 227.
Yan, Q. and de Pablo, J., 2000, *J. Chem. Phys.* **110**, 1276.
Zhou, F. and Schulten, K., 1995, *J. Phys. Chem.* **99**(7), 2194.

2
Red Blood Cell Shapes and Shape Transformations: Newtonian Mechanics of a Composite Membrane

Gerald Lim H. W., Michael Wortis, and Ranjan Mukhopadhyay

Abstract

The normal human red blood cell has at equilibrium the shape of a flattened biconcave disc about 8 μm in diameter. A variety of chemical and physical stresses cause the normal red cell to deform in a systematic and universal way to form, on the one hand, invaginated shapes called "stomatocytes" and, on the other, spiculated shapes called "echinocytes." This series of shapes is called the stomatocyte-discocyte-echinocyte or SDE sequence. It is now believed that the SDE sequence is largely controlled by a single mechanical parameter which specifies the extent to which the cell membrane prefers a concave or convex shape. This mechanism, first proposed by Sheetz and Singer in 1974, is called the bilayer–couple hypothesis. To understand the SDE sequence and to test the bilayer–couple hypothesis it is necessary to understand the structure of the red-cell membrane and to model it in terms of the variables which characterize its mechanical properties. The red-cell membrane is a composite structure consisting of a fluid-bilayer plasma membrane closely associated on the cytosolic side with an elastic protein network called the membrane skeleton. The plasma membrane resists bending but has no shear resistance. The membrane skeleton is comparatively soft but resists both stretch and shear deformation. To elucidate the shapes and shape transformations of the red cell it is necessary to understand the mechanics of these fluid and elastic membranes, both separately and in their composite state.

This article reviews red-cell mechanical properties and equilibrium membrane mechanics and then applies this material to the problem of modeling the red cell and understanding the SDE sequence. The motivation for such a review is the fact that, although the basic experimental facts have been known for more than sixty years and the membrane mechanics have been understood at a conceptual level for about thirty years, until recently it has not been possible reliably to compute cell shapes more complicated than the discocyte and, thus, to test in detail the bilayer–couple hypothesis and to reproduce the full range of SDE shapes. Although some of this work has already appeared in the professional literature, a full exposition has not been available and many of the results presented here are new.

Following an introductory section, which gives an overview of the field, Sections 2.2–2.4 review what is known about the mechanical properties of the red cell and how to model its equilibrium shape in terms of membrane mechanics. Section 2.5 describes the Monte Carlo technique we use for solving the shape equations. Detailed results for the model red-cell shapes which result from these calculations are presented in Section 2.6. Finally, in Sections 2.7 and 2.8, these predictions are compared with observations. We show that, using a reasonable choice of mechanical parameters, it is now possible to produce from membrane mechanics the full SDE sequence plus certain other unusual red-cell shapes, several of which have been observed.

The presentation is designed to be pedagogical and might be used as the basis for a minicourse on red-cell shapes and membrane mechanics. Section 2.1 summarizes the problem and the results. Section 2.2 gives biological background on red-cell structure and shape phenomenology. Section 2.3 provides background on membrane-shape energetics. Section 2.4 is a primer on equilibrium membrane-shape mechanics, including both fluid and elastic components. Additional important but more-technical background is provided in a series of linked Appendices. Appendix A summarizes what is known about the important mechanical parameters of the red cell and its membrane. Appendix B gives a symmetry-based discussion of the terms which appear in the membrane Hamiltonian, including an introduction to two-dimensional elastic theory. Appendix C provides a derivation and summary of results from differential geometry that are needed to formulate the mechanics of curved membranes in a convenient and fluid manner. Appendix D provides details of membrane-mechanical calculations for results quoted in Section 2.4 but too technical to present in the main text. An extensive bibliography is included.

2.1
Introduction

2.1.1
Overview and History

Red blood cells (RBCs), or "erythrocytes," were first observed by Anton van Leeuwenhoek in 1674, soon after the invention of the light microscope. In

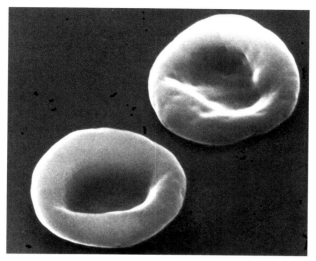

Fig. 2.1 Normal human red cells are flattened biconcave discs, a shape known as the discocyte. Copyright information: Electron micrograph courtesy of Dr. Narla Mohandas, with permission.

its normal resting state at physiological osmolarity (~ 290 mosmol/l), pH (~ 7.4) and at room temperature, the human erythrocyte assumes naturally the shape of a flattened biconcave disc (Fig. 2.1) with a diameter of 8 μm, a thickness of 1.7 μm, a volume of 90–110 μm^3 and a surface area of 130–140 μm^2 (Bessis 1973; Lichtman et al. 2005). Of course, all these numbers have a range of natural variability, in that their values vary within the RBC population of a single individual and, also, from one individual to another (more in Appendix A). An erythrocyte with this general shape is referred to as a discocyte.

In addition to its normal discocytic shape, the red cell at rest is known to assume a variety of other distinct shapes. Figures 2.2 and 2.3 illustrate some of these "unusual" erythrocyte shapes along with the terminology used to describe them. "Echinocytes" are characterized by one (or more) exterior projections or "spicules." "Stomatocytes," on the other hand, exhibit corresponding cup-shaped invaginations. A sequence of roman-numeral labels indicate the severity of the deformation. The smaller deformations, up to and including category III, occur reversibly; beyond category III, deformations are irreversible. A few other named shape categories are also shown in Fig. 2.3. Terminology in the form presently used is largely due to Bessis (1972) (Bessis 1973; Bessis 1974; Bessis 2000).

Many of these unusual shapes appear *in vivo* as infrequent anomalies in normal blood and their appearance in this context is not related to any clinical

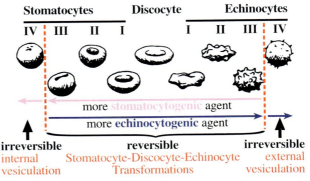

Fig. 2.2 The standard, pictorial classification of RBC shape classes in use today, originally proposed by Bessis (1972). Stage IV is irreversible, since the transformation from stage III to IV is accompanied by a loss of membrane area through vesiculation (see Section 2.2.3). Transformations among the intermediate stages are qualitatively reversible and are collectively referred to as the stomatocyte-discocyte-echinocyte shape transformations. Adapted from Fig. 2 of Betticher et al. (1995) with permission of the American Physiological Society.

Table 2.1 Shape-changing agents. Typical examples of stomatocytogenic and echinocytogenic agents. An extensive list is given in Wong (1999).

Stomatocytogenic	Echinocytogenic
Cationic amphipathic drugs	Anionic amphipathic drugs
Cholesterol depletion	Cholesterol addition
Low salt (hypotonic saline)	High salt (hypertonic saline)
Low pH	High pH
	Intracellular ATP depletion
	Proximity to glass

disease or pathology. In addition, they can be produced systematically and reliably by treating normal erthyrocytes *in vitro* with specific chemical agents. In particular, there is a broad class of agents, shown in Table 2.1, whose application drives the normal discocyte through the the stomatocyte-discocyte-echinocyte (SDE) sequence of shape transformations shown in Fig. 2.2. Application of agents in the echinocytogenic group, in sufficient concentration, drives an initially discocytic shape to the right in Fig. 2.2, while application of those from the stomatocytogenic group drives it to the left. Different agents from the same group take an initially discocytic erythrocyte through exactly the same sequence of shapes. The agents from one class counteract the agents from the other (as long as the system remains in the reversible range). Explaining the cause and mechanism of the SDE sequence, including its apparent universality, is a central focus of this paper.

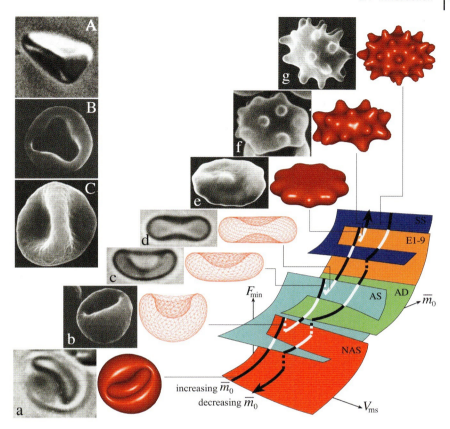

Fig. 2.3 (a–g) The SDE sequence. Side-by-side comparison of laboratory images of stomatocytes III, II and I (a–c), discocyte (d) and echinocytes I, II and III (e–g) with the corresponding calculated shapes, NAS(10), NAS(7), AS(3), AD(5), E1-9(3), SS(11) and SS(10) at $V_{\mathrm{ms}} = 148\,\mu\mathrm{m}^3$ (see Section 2.6). (A–C) Laboratory images of a non-axisymmetric discocyte, a triangular stomatocyte and a knizocyte. Plot of F_{min} as a function of V_{ms} and $\overline{m}_0$: Schematic illustration of the intersections of the surfaces of minimum energy F_{min} of the principal shape classes in the vicinity of $V_{\mathrm{ms}} = 148\,\mu\mathrm{m}^3$ (see Section 2.6.3). $\overline{m}_0$ is a dimensionless measure of the effective area difference between plasma-membrane leaflets (see Eq. (2.101)); V_{ms} measures the volume of the relaxed membrane skeleton (see Section 2.3.3 and Fig. 2.7). For clarity, we have omitted some intermediate shapes in the transformation from SS to E1-9. The directed lines show schematic trajectories through the shape classes as $\overline{m}_0$ is varied by application of echinocytogenic or stomatocytogenic agents. Dashed segments indicate sudden irreversible changes associated with hysteretic behavior.
Copyright information: Image (A) and images (c) and (d) are reprinted from Fig. 5C of Fischer et al. (1981) and Fig. 2A of Jay (1975), respectively, with permission of the Biophysical Society. Images (a), (b), (B) and (f) are reprinted from Fig. 119 of Bessis (1956), Plate I(a) and I(b) of Brailsford et al. (1980) and Fig. 1 of Jones et al. (1987), respectively, with permission of Elsevier. Images (C), (e) and (g) are reprinted from Figs. 28–16, 28–20 and 28–22, respectively, of Kimzey (1977).

It appears that Hamburger (1895) was the first to observe reversible RBC morphological transformations induced by changes of solution tonicity. Then, in the 1930s, Ponder (1948) performed comprehensive investigations, both with and without volume changes, and identified all shape classes on the discocyte–echinocyte side of the SDE sequence. Subsequent experiments in the 1960s, particularly the work of Deuticke (1968), firmly established the discocyte-stomatocyte transformations. Bessis (1972) (Bessis 1973; Bessis 1974; Bessis 2000) finally combined the discocyte-echinocyte observations with the discocyte-stomatocyte observations to form the full SDE sequence shown in Figs. 2.2 and 2.3.

In spite of the wealth of knowledge about RBC shapes accumulated over more than three centuries of observation, a full explanation of the shapes and shape transformations of the normal human RBC has remained elusive. As recently as 2001, the RBC physiologist J.F. Hoffman (2001) regarded this problem as the first of his "...own most perplexing and cherished red cell problems..." in a commentary entitled 'Questions for Red Blood Cell Physiologists to Ponder in This Millennium.' The work reported here is an answer – or at least a partial answer – to Hoffman's implied question.

Both usual and unusual erythrocyte shapes are (of course!) the result of mechanical forces operating at the microscopic level. It is the purpose of this article to describe what those forces are and how they lead to the observed shapes and shape transformations, especially the well-documented SDE sequence. We will also discuss some of the other unusual shapes. We will need as background to review the structure and composition of the red-cell membrane (Section 2.2). We will then review (Section 2.3) how the mechanical energy of this structure may be modeled in terms of a small set of material parameters, many of which are measurable *in vitro*. This energy functional, which we denote $F[S]$, depends on the shape S of the red cell. The effect of thermal fluctuations on the shape mechanics is generally – but not always – small, so the observed shapes are expected to be minima of $F[S]$ with respect to variations of S. Finding these minimizing shapes requires solving a problem in membrane mechanics (Sections 2.4 and 2.5). With these ingredients in hand we will then be able to explore in the remaining sections how the observed stationary shapes emerge from the shape mechanics and how they depend on the material-parameter inputs. Although the background material is generally well-known, much of the material presented in Sections 2.6 and 2.7 is new. The reason for this is that shape calculations for the more-complex morphologies are computationally intensive. Thus, although the key qualitative ideas have been around for several decades, it is only recently that these ideas could be validated by quantitative calculations of shapes which are not axisymmetric. A summary of results has been presented elsewhere (Wortis 2001; Lim et al. 2001; Lim et al. 2002; Mukhopadhyay et al. 2002); however, what appears here is the first published comprehensive account of our work (see also Lim (2003)).

The remainder of this section contains a summary of the overall logic of the red-cell shape problem, including the structure of the red cell, the ingredients of the shape mechanics and the form of the shape calculations. It is intended that this material should give an overview of red-cell shape mechanics at a qualitative level. The remainder of the paper fills in this framework.

2.1.2
Structure of the Erythrocyte: the Composite Membrane

Most cells have internal stress-bearing structures which determine their shapes; however, red cells are an exception to this rule. Red cells are produced from stem cells in the bone marrow. Initially, they are nucleated; however, upon exiting the marrow and entering the circulation, they lose their nucleus and additional internal structures, leaving the interior cytosol a more-or-less uniform viscous fluid, capable of supporting at equilibrium only an isotropic pressure. At this stage, they are called reticulocytes. Once in the blood stream, it takes several days for the reticulocyte to assume the normal RBC shape. After this it lives in the circulation for about 120 days and is then biochemically tagged, captured in the spleen and recycled. During its lifetime the red cell cycles (about $10^5 \times$) through the circulatory system carrying oxygen. In this process it must pass repeatedly through capillaries with diameters as small as $3\,\mu\text{m}$. In order to pass through these restrictions, the red cell deforms at constant volume and area into a sausage shape, which makes good diffusive contact with the capillary walls, and then reforms its discoid shape after the constriction. This reversible shape transformation requires energy (supplied by the heart), so there is an evolutionary advantage to building the red cell out of a soft material.

From a mechanical perspective, the red cell in circulation is effectively a bag of fluid surrounded by a quasi-two-dimensional composite membrane (Alberts et al. 2002; Lichtman et al. 2005). On the inside, the cytosol is predominately a concentrated aqueous solution of hemoglobin, but also containing various salts and biologically active proteins. The surrounding membrane is a thin composite structure consisting of two closely coupled components, which will be described in more detail in Section 2.2. The outer layer, called the plasma membrane (pm), is a self-assembled amphipathic bilayer in the fluid (L_α) phase. Its dominant ingredient is a multi-component lipid mixture; however, there is a significant admixture of proteins and other biomolecules. The plasma membrane has a thickness $D \approx 4\,\text{nm}$. It is permeable to water but effectively impermeable to ions and large molecules, so it forms a semipermeable osmotic barrier between the cytosol and the extracellular fluid. The inner layer, called the membrane skeleton (ms) or, by some authors, the cytoskeleton, is a loose elastic network of polymerized proteins which is anchored at discrete locations to the plasma membrane and extends up to 50 nm into the cytosol. Note that the membrane-skeleton thickness, although significantly larger than D, is nonetheless still small on the scale of the whole cell. Both the plasma membrane and the membrane skeleton have structure at the submicron level (see Section 2.2) but are uniform and isotropic to a good approximation on the length scale of the whole cell. Being fluid, the plasma membrane is easily deformable but rather fragile. The role of the membrane skeleton appears to be that of toughening the cell-membrane capsule so that it is not breached during the repeated large shape deformations which the red cell undergoes in the circulation.

2.1.3
What Fixes the Area and Volume of the Red Cell? Flaccid vs. Turgid Cells

The surface area A and volume V of the red cell play a crucial role in RBC morphology. In all our computations we have used the values $A_0 = 140\,\mu\text{m}^2$ and $V_0 = 100\,\mu\text{m}^3$, as shown in Table 2.2 (see also Appendix A). It will be useful in what follows to define a characteristic length R_A by

$$A_0 \equiv 4\pi R_A^2, \qquad (2.1)$$

so the volume of a sphere of area A_0 is

$$V_{A_0} = \frac{4\pi}{3} R_A^3. \qquad (2.2)$$

Based on the area given above, one finds $R_A = 3.34\,\mu\text{m}$, which is an important length scale in the red-cell shape problem. The dimensionless ratio,

$$v = \frac{V_0}{V_{A_0}}, \qquad (2.3)$$

is called the reduced volume. For the red cell $v = 0.642$, as tabulated in Table 2.2.

The red-cell area is effectively fixed at $A = A_0 = A_{RBC}$ by the amount of material in the plasma membrane. Of course, in principle, the area responds elastically to the tension created by any pressure difference ΔP across the membrane; however, the area expansion modulus of the plasma membrane $K_A \approx 0.5\,\text{J}/\text{m}^2$ (Katnik and Waugh 1990) is large enough so that, for typical pressure differences, any area change is negligible.[1]

---

1) The typical pressure difference for a flaccid RBC is $\Delta P \sim 5 \times 10^{-3}\,\text{J}/\text{m}^2$, which translates to a fractional area change of roughly 2×10^{-8}. This estimate is based on a membrane energy $E \sim \kappa_b$ with $\Delta P \sim \frac{dE}{dV} \sim \frac{\kappa_b}{R_A^3} \sim \frac{2\tau}{R_A}$ with $\tau = K_A \frac{\Delta A}{A_0}$, where τ is the membrane tension (see more at Sections 2.3 and 2.4 and Appendix A).

2.1 Introduction

Table 2.2 Parameters of the model. Defining equations and further references are given at the right.

Cell and membrane geometry		
Red-cell area	$A_0 = A_{RBC} = 140\ \mu m^2$	Appendix A
Red-cell volume	$V_0 = V_{RBC} = 100\ \mu m^3$	Appendix A
Scale length	$R_A = 3.34\ \mu m$	Eq. (2.1)
Maximum volume of red cell	$V_{A_0} = 4\pi R_A^3/3 = 155.8\ \mu m^3$	Eq. (2.2)
Reduced volume of red cell	$v = V_0/V_{A_0} = 0.642$	Eq. (2.3)
Bilayer thickness	$D = 4\ nm$	
Offset of leaflet midplanes	$D_0 \approx 2\ nm$	Eq. (2.21)
Membrane skeleton thickness	$D_{ms} \approx 50\ nm$	
Constraint moduli		
Area modulus of pm	$K_A = 0.5\ J/m^2 = 2.5 \times 10^6\ \kappa_b/\mu m^2$	Eq. (2.9)
Osmotic modulus	$K_V = RTc_0$	Eq (2.12)
	$= 7.23 \times 10^5\ J/m^3 = 3.61 \times 10^6\ \kappa_b/\mu m^3$	
Computational area modulus	$K_A^* = 1.0 \times 10^{-3}\ J/m^2 = 5 \times 10^4\ \kappa_b/\mu m^2$	Eq. (2.83)
Computational osmotic modulus	$K_V^* = 1.0 \times 10^4\ J/m^3 = 5 \times 10^4\ \kappa_b/\mu m^3$	Eq. (2.84)
Plasma-membrane parameters		
Bending modulus	$\kappa_b = 2.0 \times 10^{-19}\ J/m^2 \sim 48\ k_B T_{room}$	Eq. (2.14), Appendix A
Gaussian bending modulus	$-2\kappa_b \leq \kappa_g \leq 0$ (unknown, Appendices B.2, D.3)	Eq. (2.15), Appendix A
Area-difference modulus	$\bar{\kappa} = 2\kappa_b/\pi = 1.27 \times 10^{-19}\ J$	Eq. (2.16), Appendix A
Ratio of moduli	$\alpha_b = \bar{\kappa}/\kappa_b = 2/\pi = 0.637$	
Spontaneous curvature	$C_0\quad (c_0 = R_A C_0)$	Eq. (2.14)
Area difference (relaxed)	$\Delta A_0\quad (\Delta a_0 = \Delta A_0/D_0 R_A)$	Eq. (2.16)
Area difference (actual)	$\Delta A\quad (\Delta a = \Delta A/D_0 R_A)$	Eq. (2.21)
Effective C_0	$\bar{C}_0\quad (\bar{c}_0 = R_A \bar{C}_0)$	Eq. (2.23)
Area-difference parameter $\bar{m}_0$	$\bar{m}_0 = 2\bar{c}_0/\alpha_b\quad (-60 < \bar{m}_0 < 160)$	Eqs. (2.33), (2.101)
Membrane-Skeleton parameters		
Stretch modulus of ms	$K_\alpha = 5 \times 10^{-6}\ J/m^2 = 25\ \kappa_b/\mu m^2$	Eq. (2.27)
Shear modulus of ms	$\mu = 2.5 \times 10^{-6}\ J/m^2 = 12.5\ \kappa_b/\mu m^2$	Eq. (2.27)
Elastic length scale	$\Lambda_{el} \sim 0.3\ \mu m$	Eq. (2.5)
Nonlinear elastic coefficients	$a_3 = -2, a_4 = 8, b_1 = 0.7, b_2 = 0.75$	Eq. (2.28)
Relaxed area of ms	$A[S_0] = A_0 = 140\ \mu m^2$	Fig. 2.7
Relaxed volume of ms	$V_{ms} = V[S_0]\quad (V_0 \leq V_{ms} \leq V_{A_0})$	Fig. 2.7

The red-cell volume V, on the other hand, is set by osmotic equilibrium. The osmotic pressure of any significant osmotic imbalance is almost always appreciably larger than any pressure difference ΔP across the membrane,[2] so water flows through the membrane until the tonicities of the cytosol and

---

[2] According to the van't Hoff relation $\Delta P = RT\Delta c$, where R is the gas constant and Δc is the tonicity difference across the membrane in moles/m^3. Thus, at the physiological osmolarity of 290 mosmol/l, a typical flaccid-vesicle pressure difference of $\Delta P \sim 3 \times 10^{-3}\ J/m^2$ is supported by a fractional difference in tonicity of $\frac{\Delta c}{c} = \frac{\Delta P}{RTc} \sim 7 \times 10^{-9}$.

the extracellular solution are effectively equalized. Thus, if a red cell is placed in hypertonic solution, water flows outwards from the cytosol, deflating the cell. Conversely, if the extra-cellular solution is hypotonic, water flows inwards, inflating the cell. Depending on the degree of hypotonicity, one of three things happens. If the original osmotic imbalance is sufficiently small, then inflation stops (i.e., osmotic balance is achieved) at $V < V_{A_0}$, the maximum volume that can be accommodated by the original, unstretched membrane area A_0. In this range the pressure difference ΔP remains small[1] and the resulting vesicle is said to be "flaccid." For larger osmotic imbalance, the in-flowing water begins to push against the membrane-area limitation, thus forcing the area to increase above A_0 and bringing the expansion modulus into play. In this situation the membrane tension τ and the pressure difference ΔP both increase rapidly, and the vesicle is said to be "turgid." The shape of a turgid vesicle is spherical to a good approximation. Finally, for still higher osmotic imbalances, the membrane tension exceeds the lysis tension and the plasma membrane ruptures.[3] At this point cytosol escapes, relieving the excess pressure and generally removing some osmogenic solutes. With the pressure relieved, the line tension at the broken edge can lead to resealing of the plasma membrane and osmotic flow resumes. This cycle of lysis followed by resealing can be repeated several times, until the red cell is left finally in osmotic balance in either the flaccid or the turgid state (although normally with a damaged membrane skeleton).

The *in vivo* red cell, which finds itself subject to physiological osmolarity, has biochemical feedback loops (Strange 1994; Lang 2006) which adjust its cytosolic tonicity to keep its volume at $V_0 = V_{RBC} \approx 0.642\, V_{A_0}$, in the flaccid range. When such a cell is placed *in vitro* in a solution of non-physiological tonicity, it swells or deflates on a time-scale faster than the internal feedback. In this article we will be concerned principally with the shapes of undamaged flaccid red cells. In many – but not all – experiments, the extra-cellular solution is maintained at physiological tonicity, so the cell volume remains at V_0. Work involving the shapes of cells subject to osmotic inflation ("sphering") and/or deflation has also been carried out (Furchgott and Ponder 1940; Ponder 1948; Rand and Burton 1963; Rand 1967; Fung and Tong 1968; Canham 1970; Canham and Parkinson 1970; Skalak et al. 1973; Zarda et al. 1977; Pai and Weymann 1980) but is not specifically addressed in this work.

---

3) Lysis occurs when the area has increased by only a percent or two, that is, at a tension of $\tau \sim K_A \frac{\Delta A}{A_0} \sim \frac{K_A}{100} \sim 5 \times 10^{-3}\,\mathrm{J/m^2}$. At this point the pressure difference has increased by a factor of roughly 6×10^5 above its flaccid value. However, even at this higher value, the corresponding osmolarity difference is still small, $\frac{\Delta c}{c} \sim 4 \times 10^{-3}$.

2.1.4
Shape Determination for a Flaccid Red Cell at Equilibrium: Membrane-Energy Minimization

What determines the shape of a flaccid red cell at equilibrium in aqueous solution? Red cells do not contain internal force-bearing structures to regulate their shapes. Flaccid red cells have $V_0 < V_{A_0}$, so they are not maintained spherical by internal pressure. The cell membrane is surrounded only by fluids which, on the scale of the cell size R_A, are uniform and isotropic. It follows that the cell shape can only be determined by the shape preference of the cell membrane itself. Like any other mechanical system in a dissipative environment (the fluids inside and outside) the membrane will tend towards its state of lowest mechanical energy. Thus, if $F[S]$ is the membrane energy (strictly speaking, its free energy) as a functional of the cell-membrane shape S, the observed shape at mechanical equilibrium will be one which minimizes F with respect to changes in S, subject to the constraints that the cell volume and area have at their prescribed values. Energy minimization implies that the equations of mechanical equilibrium are satisfied: cells take on shapes determined by membrane mechanics.

Clearly, a central ingredient of equilibrium-shape determination is the form of the shape-energy functional $F[S]$, which we discuss in the next subsection and which we will explore in full detail in Section 2.3. Once $F[S]$ is known, finding the equilibrium shape is reduced to the computational problem of finding a constrained energy minimum. $F[S]$ depends on certain material parameters or moduli, whose values control the equilibrium shapes.

There are two factors which will make this picture somewhat more complex. One is that, under given conditions, there may be more than one mechanically stable state. The second is that thermal fluctuations are always present. Except in special cases of degeneracy, one of the stable states – the "ground state" – has the lowest energy, while any others are metastable. If the energy $k_B T$ of thermal fluctuations were large on the scale of the membrane-energy landscape, then what would be seen in the lab would be a fluctuating thermal ensemble of all the states – equilibrium and non-equilibrium – within roughly $k_B T$ of the ground state. It will, however, turn out that the membrane energy scale is set by the bending modulus of the plasma membrane $\kappa_b \approx 2 \times 10^{-19}$ J $\sim 50 \, \kappa_b T_{\text{room}}$ (see Section 2.3 and Appendix A). In this sense, red-cell shape is a low-temperature problem, so thermal fluctuation effects are normally small, although there are exceptions to this rule near mechanical instabilities. At the same time, it means that barriers between the ground state and any metastable states will generally be high, so that we may anticipate – and will encounter – metastability and hysteresis in the cell-shape mechanics.

2.1.5
Ingredients of the Membrane Shape-Energy Functional $F[S]$

The shape-energy functional $F[S]$ will be described in detail in Section 2.3; however, it will help our overview to give here a qualitative account of its most important ingredients. We saw in Section 2.1.2 that the membrane is composite. Correspondingly, we will write the membrane energy as a sum,

$$F[S] = F_{\rm pm}[S] + F_{\rm ms}[S], \qquad (2.4)$$

referring to the plasma membrane (pm) and the membrane skeleton (ms), respectively. $F_{\rm pm}[S]$ encodes the mechanics of the bilayer. Since the plasma membrane is fluid, it cannot support shear stress. Thus, its mechanics is dominated by its resistance to bending, which is characterized by the bending modulus $\kappa_{\rm b}$ and by a preferred mean curvature of the membrane surface, denoted $\frac{1}{2}\overline{C}_0$, which reflects the asymmetry between the inner and outer leaflets of the bilayer. The quantity $\overline{C}_0$, called the effective spontaneous curvature, is defined (see Section 2.3.2) in such a way that $\overline{C}_0 > 0$ promotes outward curvature, that is, spiculated shapes like echinocytes, and $\overline{C}_0 < 0$ promotes inward curvature, that is, invaginated shapes like stomatocytes. $F_{\rm ms}$ encodes the elasticity of the cytoskeletal protein network, characterized by stretch and shear moduli K_α and μ, respectively, with $K_\alpha \approx 2\mu \sim 5 \times 10^{-6}\,{\rm J/m^2}$.[4] To define fully the energy of stretch and shear, it is necessary to know the strain and energy of some reference state of the membrane skeleton. We shall take this state to be a hypothetical "relaxed", that is, unstressed, reference shape S_0. Thus, the energy of the membrane skeleton depends functionally on both S and S_0, and we shall write it as $F_{\rm ms}[S_0; S]$ when we wish to emphasize this dependence. It would be simple to assume that S_0 is simply a sphere of area A_0; however, it will turn out (Sections 2.6 and 2.7) that there is good reason to believe that this is not so and that, instead, S_0 is somewhat oblate.

The relative importance of the two terms in Eq. (2.4) is measured by the ratio of the bending modulus to, say, the shear modulus, which defines an "elastic length scale" Λ_{el} (Mukhopadhyay et al. 2002), according to

$$\Lambda_{el}^2 \equiv \frac{\kappa_{\rm b}}{\mu}, \qquad (2.5)$$

giving $\Lambda_{el} \sim 0.3\,\mu{\rm m}$. Thus, it follows on dimensional grounds that, when elastic stresses in the membrane skeleton are present, they will tend to domi-

---

[4] Note that K_α (for the membrane skeleton) is five orders of magnitude smaller than K_A (for plasma membrane). Thus, in a uniform stretching of the composite membrane, the effect of K_α would be entirely negligible. Indeed, it is the large value of K_A that holds the cell-membrane area effectively fixed. However, as the cytoskeleton redistributes itself over the plasma membrane to minimize the total energy, the strain in the cytoskeleton is by no means uniform, and the contribution of $F_{\rm ms}$ (including the stretching term) to the total energy can be locally important, as explained below. The failure to recognize the quite different roles played by K_α and K_A has lead to some confusion in the literature.

nate the shape problem at length scales larger than Λ_{el}, while at length scales smaller than Λ_{el} the shape will be controlled by the bending energy. Note that $R_A > \Lambda_{el} \gg D$, so the elastic length scale is smaller than the cellular scale but still much larger than the membrane thickness. The upshot for the red cell is that, for relatively smooth shapes like the discocyte, the membrane skeleton is not subject to significant stress (and certainly not to significant stress inhomogeneity), so $F_{ms}[S]$ does not play a very important role in shape determination. On the other hand, for the more pronounced spiculated or invaginated shapes, appreciable cytoskeletal stresses are present. In this situation, we may expect that features of cell shape on scales smaller than Λ_{el} are controlled by the bending energy $F_{pm}[S]$ (the size of which is measured by κ_b), while those on scales larger than Λ_{el} are dominated by the energy cost of cytoskeletal deformations, as encoded in $F_{ms}[S]$ (the size of which is measured by K_α and μ). As a consequence we will find, for example, that the typical spicule size is comparable to Λ_{el}.

2.1.6
Shape Classes, Stability Boundaries and Phase Diagrams

The full shape-energy functional is, as we have seen above, a functional of the membrane shape S, the unstressed cytoskeletal shape S_0, the cell volume and area, and a number of material parameters $\{p_i\}$ (such as the elastic moduli), that is, $F[A_0, V_0, \{p_i\}, S_0; S]$. To find the mechanically stable shape or shapes for given $A_0, V_0, \{p_i\}, S_0$, we will minimize over S to produce a set of shapes $S_{\min}^{(\alpha)}[A_0, V_0, \{p_i\}, S_0]$ with corresponding energies $F_{\min}^{(\alpha)}[A_0, V_0, \{p_i\}, S_0]$. If one (or more) of the parameters inside the brackets varies smoothly, the corresponding minimizing shapes and energies will also, in general, vary in a smooth manner. Thus, each minimizing shape $S_{\min}^{(\alpha)}$ becomes a class of shapes continuously related to one another and, correspondingly, each energy $F_{\min}^{(\alpha)}$ becomes a "sheet" or "branch" over the "phase space" $[A_0, V_0, \{p_i\}, S_0]$, as illustrated in Fig. 2.3. Each of these shape classes (and each of the corresponding energy sheets) carries a label α. We will choose this shape-class label descriptively, and in some (but not all) important cases the classes will be associated with simple symmetries. Thus, in Fig. 2.3, the label AD for "axisymmetric discocyte" refers to shapes similar to the normal discocyte, with a rotation axis plus an up/down reflection plane perpendicular to that axis. Similarly, the "axisymmetric stomatocyte" (AS) class has axisymmetry but lacks the up/down reflection plane. Other classes will be described and discussed in Section 2.6.1.

Characteristically, each class of stable equilibrium shapes exists over some limited range of the phase-space variables. These limits correspond often (but not always) to mechanical instabilities, and they are usually associated with mathematical bifurcations. We will, therefore, refer both to the edges of the sheets and to their projections onto the underlying phase space $[A_0, V_0, \{p_i\}, S_0]$ as stability boundaries. Beyond the stability boundaries of a particular class, stable shapes in that class do not exist.[5] The relationship of distinct sheets to one another depends on the details of the bifurcation structures. Sheets may pass through one another, when changing the value of a parameter causes the relative energy of two sheets of distinct shapes to interchange. Alternatively, two sheets can meet along a stability boundary as, for example, when a sheet of lower symmetry bifurcates from a sheet of higher symmetry. This will be discussed at greater length in Sections 2.6 and 2.7.

The loci in phase space of all stability boundaries and sheet intersections constitute a kind of generalized phase diagram. We will wish to refer to some particular subsets of this generalized phase diagram (see Fig. 2.4). The first and simplest such subsets are the stability diagrams, which show for each shape class α the outline of the region of phase space over which a shape of that class is mechanically stable. A stability diagram is, of course, just the locus of the stability boundaries surrounding the class α. Up to now we have said nothing about the relative energies of the various sheets. This is relevant, since, at the longest time scales and at low-enough temperatures, all cells find their lowest-energy (ground) state. The "$T = 0$ phase diagram" or "shape diagram" is a map over phase space of the regions where each shape class contains the ground-state shape. The shape diagram contains two kinds of boundaries (see Fig. 2.4). The first are boundaries where two sheets cross one another (B and C in the generic example of Fig. 2.4). This happens in regions of the phase diagram where there are two (or more) distinct stable shapes and the energy sheets of the two lowest-energy shape classes intersect. On one side of the intersection, one shape class has the lowest energy; on the other side, the order is interchanged. The loci of these lowest-energy intersections we will refer to as "discontinuous" or "first-order" phase boundaries, since the lowest-energy shape changes abruptly across such boundaries.[6] Alternatively, a ground-state region may be bounded by a stability boundary where, at a bifurcation, one shape class becomes smoothly unstable to another (A and B in the generic example of Fig. 2.4). Such boundaries we refer to as "continuous" or "second-order," since one class of shapes

5) Sometimes a sheet continues beyond its stability boundary as a sheet of unstable equilibrium shapes. Such mechanically unstable shapes are not observable in the lab, nor are they picked up computationally by the kind of energy-minimization process described in Section 2.5, although they would show up as solutions of the equations of mechanical equilibrium (Section 2.4).
6) The terminology is suggested by that of thermodynamic phase transitions. However, it is important to keep in mind that sharp transitions in the shape problem are strictly a zero-temperature phenomenon. The red cell is a finite system; thus, all sharp boundaries disappear at any $T > 0$ and all that remains is a smoothly changing thermodynamic shape ensemble.

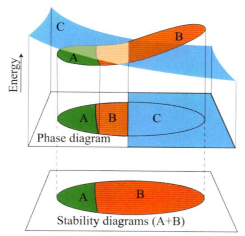

Fig. 2.4 Schematic representation of energy sheets with associated phase and stability diagrams. Energy is measured in the vertical direction; the horizontal coordinates represent control parameters. There are three shape classes, A, B and C, in this example. The transition between A and B is continuous, so the energy sheets for these two classes join smoothly. An independent sheet C intersects B and forms the lowest-energy (ground) state on the right-hand side. The black lines outlining the A+B sheet are stability boundaries; the phase C is assumed stable over the entire region. The generalized phase diagram shows the phase and stability as projected onto the base plane. The B/C phase boundary is discontinuous; the A/B phase boundary is continuous. The joint stability diagrams of the A+B phases are shown below. Note that B remains locally stable beyond the B/C intersection, so that moving to the right on the sheet B through the intersection will not generally result in an abrupt shape change until and unless the energy barrier between the B and C configurations becomes comparable to the thermal energy $k_B T$. In the uncolored region of the phase diagram, some other shape class or classes (not shown) must be present, since the ground-state energy always changes continuously.

blends smoothly into another class with lower symmetry. Because the shape changes smoothly across a second-order boundary, such transitions may be hard to locate experimentally with precision, especially when thermal fluctuations are significant.

To appreciate the significance of the shape diagram, consider a laboratory experiment in which one (or more) control parameters (e.g. $V_0, \overline{C}_0, \ldots$) are varied smoothly, tracing out a continuous one-dimensional trajectory in phase space (see Fig. 2.3). If the change is carried out too fast, equilibrium shapes are not observed, because the system does not have time to equilibrate. On the other hand, if it is carried out very slowly, then the system has time to find its ground state, so the shape classes observed will be those of the shape diagram. There is, however, an intermediate situation, where the shape can equilibrate to a stable state but not necessarily find the ground state due to the generically high energy barriers between distinct energy minima (see Section 2.1.4). This situation is actually quite common. In this case, second-order boundaries remain sharp in principle but, as always, difficult to locate precisely in experiment; however, first-order boundaries typically become moot in experiments because of hysteretic effects. Thus, the observed shape will typically continue on what was the ground-state sheet beyond the nominal first-order transition point into the the metastable region and drop to a lower-energy sheet only as the system approaches a stability boundary, where the energy barriers become comparable to the thermal energy $k_B T$.

In Section 2.6, we will calculate both stability diagrams and shape (phase) diagrams for the red-cell problem. However, the calculations are time-consuming, and we so will only be able to explore a limited region of the full phase space.

2.1.7
Understanding the SDE Transformation Sequence: Universality and the Bilayer–Couple Hypothesis

The key feature of the SDE transformations is that they are "universal" in the sense that the different agents in each of the two classes listed in Table 2.1 drive the red cell through the same sequence of shapes. This universality suggests that there is a single dominant mechanism or parameter which controls the RBC shape and that the different inducing agents all "turn the same knob." The identity of this dominant driving force was correctly surmised by Sheetz and Singer (1974), although at the time there was no quantitative theory of red-cell shapes, so their insight could not be tested by a comparison of calculations with observation. Indeed, the first comprehensive and quantitative test of Sheetz-Singer is provided by the present work (Lim et al. 2002).

Their idea, called the "bilayer–couple hypothesis," was that, because of the spatial offset, $D_0 \approx D/2$, between the midplanes of the two leaflets of the plasma membrane, any difference in area $\Delta A_0 = A_{out} - A_{in}$ between the two leaflets serves to produce a bending moment or couple.[7] If the outer leaflet is larger, then that couple will tend to make the membrane bow outward into spicules. If the inner leaflet is larger, then the couple will tend to produce inward curvature, that is, invagination. Thus, any process which increases the number of molecules or the area per molecule in the outer leaflet will produce a positive ΔA_0 and a tendency towards spiculation. Conversely, increasing the number or area of inner-leaflet molecules will produce a neg-

---

7) Of course, such a couple could not be maintained if the two leaflets exchanged lipid molecules freely, since it would immediately be relaxed by lipid "flip-flop." We shall find in Section 2.2.1 that flip-flop is inhibited and, indeed, that cellular metabolic processes control specifically and individually the lipid composition of the two leaflets.

ative ΔA_0 and invagination. We will discuss in Section 2.8.4 how each of the agents in the left column of Table 2.1 might produce a negative ΔA_0 and each of the agents in the right column, a positive ΔA_0. It is necessary to understand how this quantity ΔA_0 enters the shape mechanics described in Section 2.1.5. What we will show in Section 2.3.2 is that there is a linear relation between the effective spontaneous curvature $\overline{C}_0$ and the area difference ΔA_0.

While the bilayer–couple hypothesis is, we believe, correct, the full story has an extra complication, not appreciated until considerably later than the seminal paper of Sheetz and Singer (1974). The bilayer–couple refers only to the bilayer component of the red-cell membrane. In this spirit there was considerable work starting with Helfrich and coworkers (Helfrich 1973; Deuling and Helfrich 1976) and continuing into the 1990s which studied the equilibrium shapes of *in vitro* fluid-phase lipid-bilayer vesicles, based on the energy functional $F_{\rm pm}[S]$ (only), in the expectation that vesicles might be good models for red-cell shape. We will review this work in Section 2.3.5; however, suffice it to say for the present that this work was only partially successful. While it turns out that appropriate neutral and negative values of $\overline{C}_0$ do produce credible discocytic and stomatocytic shapes, echinocyte vesicle shapes are never seen. Instead, for positive values of $\overline{C}_0$, vesicles first become pear shaped and then tend to "vesiculate," that is, to form small spherical or nearly spherical "buds" with dimensions comparable to $1/\overline{C}_0$ and connected to the main body of the vesicle by a narrow neck (Miao et al. 1991; Seifert et al. 1991; Fourcade et al. 1994; Miao et al. 1994). Such shapes are not characteristic of the intact red cell. It was eventually recognized by the community (Waugh 1996; Khodadad et al. 1996; Iglič 1997; Iglič et al. 1998a; Iglič et al. 1998b; Wortis 1998) that the cytoskeletal contribution $F_{\rm ms}[S]$ solves this problem by assigning high shear energy to the narrow neck of the vesiculated shapes, thus converting them to echinocytic shapes, as illustrated in Fig. 2.5. Turning this recognition into a calculation was not easy, since at the time only

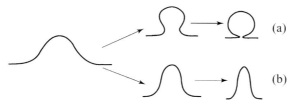

Fig. 2.5 Effect of the membrane skeleton for large values of $\overline{C}_0$. In the absence of cytoskeleton (a), increasing the value of $\overline{C}_0$ leads to the formation of buds of characteristic radius $1/\overline{C}_0$ connected to the rest of the cell by a narrow neck. In the presence of a membrane skeleton (b), the neck region experiences high shear, so budding is replaced by spicule formation.

axisymmetric shapes could be calculated. The first individual spicule shapes were calculated by Waugh (1996), Iglič (1997), Iglič et al. (1998a), Iglič et al. (1998b) and by our group (Mukhopadhyay et al. 2002) on the basis of various simplifying assumptions; however, not until this work (Lim et al. 2002) has it become feasible numerically to calculate full spiculated shapes.

2.1.8
Perspective and Outline

After reviewing red-cell components and shape observations in Section 2.2 and models of membrane structure in Section 2.3, Section 2.4 is devoted to a review of the equations of shape mechanics of first the pure-lipid plasma membrane and then the full composite plasma membrane-plus-membrane skeleton. Although these equations are useful background, they have not proved to be practical tools in calculation except in situations with axisymmetry. Section 2.5 explains the method of numerical calculation that we have used to find red-cell shapes. This consists in representing the composite membrane configuration as a two-dimensional mesh, developing algorithms for calculating the mesh energy (including both bending and stretch/shear elasticity) and then using Monte Carlo methods to find shapes that are local energy minima.

Our results for red-cell shapes and phase diagrams are presented in Sections 2.6 and 2.7. They focus principally on two axes of the high-dimensional phase space, the parameter $\overline{C}_0$ (effectively, the Sheetz-Singer parameter) and a parameter V_{ms} labeling the sphericity of the relaxed membrane-skeleton shape S_0. All the various shapes in the SDE sequence (Fig. 2.2) do appear in the shape (phase) diagram, along with several other unusual shapes, some of which are seen in experiments. This result – although encouraging – is a necessary but far from sufficient condition for "success." To validate the theory completely it would be necessary to show that, at each point in phase space, the calculated shape agrees with what is seen in the lab. Unfortunately, this kind of validation is impossible, since several of the phase-space parameters are known only poorly (see Appendix A) and some of them (like the cytoskeletal shape) are not known at all. Thus, our objective must remain more modest.

The key test that we will focus on is the sequence of predicted vs. observed shapes as the Sheetz–Singer parameter is varied. What we will find in Section 2.7.2 is that – for the parameter choices we adopt (Table 2.2 and more in Appendix A) – the observed SDE sequence of shape transformations does emerge from the minimization problem but only for a rather narrow range of cytoskeletal shapes S_0, close to the spherical limit but still appreciably oblate. Outside of this range some members of the sequence are missing or other different shape classes (e.g., Fig. 2.3) occur. Section 2.7 summarizes the principle results and predictions of our model.

Overall, the success of the calculations presented here in reproducing the observed shapes and shape changes suggests that – despite its simplicity – the model captures the key features of the red-cell mechanics and, also, that the parameters we have put in are reasonable. The limitations of the model and the robustness of the parameter choices are discussed further in Section 2.8.

What is the significance of these conclusions from the biological point of view? In a sense it would seem that we have come full circle. Initially it might have been thought that the complex red-cell shape changes reflected subtle biochemical control mechanisms. Instead, it now appears that they arise from Newton's laws (membrane mechanics) and material properties (moduli), driven dominantly by a single biochemically-controlled parameter ($\overline{C}_0$). Indeed, it is precisely the universality of the SDE transformations that makes them – in the end – rather uninteresting from the biochemical perspective. Many different inputs produce identical outputs, so the observation of output gives us only a crude, one-dimensional picture of the biochemical input. Biological-physics problems often (usually?) turn out to be more "biological" than "physical" in the sense that they depend importantly on complex biochemical, DNA-mediated control mechanisms. By contrast, the red-cell shape turns out to fit the "physics" paradigm of complex behavior arising from simple inputs. We must immediately add that the biochemical processes that influence or control $\overline{C}_0$ are complicated and interesting. And, it may well turn out that biological complexity makes itself felt at other length and time scales (see Section 2.8.5).

Once the shape mechanics are fully understood (some further fine-tuning of parameters may still be required, see Section 2.8), the way will be open to inverting the logic and using shape mechanics as a probe of the biochemical parameters entering $\overline{C}_0$. Thus, we anticipate that shape observations will be used in the future to infer the bilayer area difference ΔA_0. Watching changes in ΔA_0 will allow quantitative monitoring of changes in the number of molecules (or of the area per molecule) in each leaflet of the plasma membrane. At this stage it begins to become feasible to quantitate the effects of inducing agents such as those listed in Table 2.1 and, thus, to probe the corresponding biochemical processes via physical shape observations.

Overall this article has two purposes. On the one hand, it contains a review of the mechanical structure of the red-cell membrane and the models that have been proposed for the calculation of red-cell shapes from first principles. On the other hand, most of the results given for red-cell shapes and shape transformations are reported here for the first time, so this article is also a monograph on a particular approach to the predictive understanding of red-cell shapes.

2.2
Structure of the Cell Membrane; the SDE Sequence

In this section, we will review the structure of the RBC membrane, starting with the plasma membrane (Section 2.2.1) and continuing to the membrane skeleton (Section 2.2.2). We conclude by giving additional background on the SDE sequence and other observed red-cell shapes (Section 2.2.3).

The structural organization of the the RBC membrane is sketched in Fig. 2.6. Note that the detail is taken from the lower surface of the red cell, so the membrane skeleton, which is on the inside, appears on top of the plasma membrane. The plasma membrane is the bilayer sandwich shown schematically; the membrane skeleton is the linear protein network just inside the plasma membrane. The membrane skeleton is coupled to the plasma membrane via protein complexes which are anchored into the plasma membrane via hydrophobic domains. The plasma membrane is a 2D fluid, so the anchoring complexes are free to move in the plane of the membrane as the skeleton adjusts to lower its elastic energy. In what follows, we will give a brief characterization of these two parts of the membrane. Good general references on red-cell membrane structure are Steck (1989), Mouritsen and Andersen (1998) and Lichtman et al. (2005).

Overall, the red-cell membrane is composed of a mixture of lipids and proteins plus a small amount of carbohydrate. The lipids are confined to the plasma membrane and dominate its structure (see Section 2.2.1). A majority of the protein component is associated with the membrane skeleton and is peripheral to the membrane. The remaining minority fraction associated with the plasma membrane consists mainly of integral membrane proteins, anchored into the plasma membrane by one or more hydrophobic α-helical bilayer-spanning domains.

2.2.1
Plasma Membrane

The plasma membrane is a self-assembled bilayer in the fluid (L_α) phase. Self-assembly is driven by the amphipathic structure of the lipids, which combines a polar headgroup with one or more (usually two) hydrophobic hydrocarbon chains. The lipids are highly diverse. The dominant species (with approximate weight percents (Alberts et al. 2002)) are cholesterol (Chol, 23%), phosphatidylethanolamine (PE, 18%), sphingomyelin (SM, 18%), phosphatidylcholine (PC, 17%), phosphatidylserine (PS, 7%), glycolipids

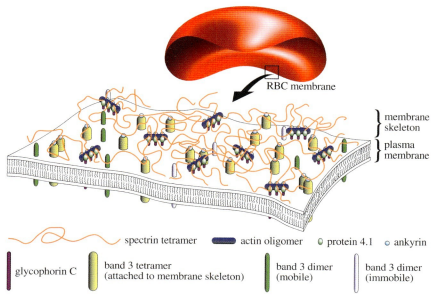

Fig. 2.6 Schematic view of the structural organization of the RBC membrane. As drawn, the interior of the cell lies above the plasma membrane, so the cytoplasm bathes the cytoskeletal protein layer (mainly spectrin) which forms the membrane skeleton. Note that the membrane skeleton is quasi-two-dimensional and does not extend significantly into the cell interior, which is otherwise filled by a concentrated solution of hemoglobin. After Fig. 1 of Schmidt et al. (1993).

(3%) and others (13%). These materials are not distributed symmetrically between the two leaves of the bilayer. Most of the PC, SM and glycolipids are in the outer leaflet, while most of the PE and PS are in the inner leaflet. At physiological pH most of these species are neutral (typically zwitterionic) except for PS, which carries a net negative charge in solution. Thus, the outer leaflet is electrically neutral, while the presence of the PS gives the inner leaflet a negative charge (which is, of course, neutralized by a nearby positively charged counterion layer). This lipid asymmetry is believed to be maintained actively in the functioning erythrocyte by a variety of specific ATP-dependent enzymes called lipid translocases. If the cell is deprived of ATP, lipid redistribution occurs, presumably towards a more symmetrical state; however, this passive "flip-flop" – thermally activated exchange between leaflets – is a slow process for most lipids because of the high energy required to make the polar headgroup pass through the region of hydrophobic tails. By contrast, because of its small polar head group, cholesterol flips relatively easily between leaflets. Under normal conditions cholesterol partitions with a mild preference for the outer leaflet (Lange and Slayton 1982; Steck et al. 2002). Both the active and passive flip-flop processes are typically slow on the scale of 10^{-1}–1 s required for mechanical shape changes. For this reason, the area difference ΔA_0 and the spontaneous curvature C_0 (the two factors which enter $\overline{C}_0$, see Section 2.3.2) are for practical purposes fixed over the time-scales of shape-changing events. By the same token, lipid flip-flop can be a mechanism for shape relaxation at longer time scales. All these considerations will become relevant when we return in Sections 2.8.4 and 2.8.5 to consider the mechanisms by which the various shape-changing agents of Table 2.1 act.

In the composite red-cell membrane lipids and proteins are in a ratio of roughly $3 : 4$ by weight (Gennis 1989), so proteins are a significant component. The non-cytoskeletal membrane proteins are very diverse, including both structural proteins, like the band 3 tetramer, the protein 4.1, the ankyrin, the actin oligamer and glycophorin, which cross-link the membrane skeleton and anchor it locally to the plasma membrane (see Fig. 2.6), and a hosts of other functional proteins, including pores, channels, transporters, signaling complexes and so forth.

2.2.2
Membrane Skeleton

The membrane skeleton is a quasi-triangular protein network composed mainly of spectrin, actin and band 4.1 plus the transmembrane anchoring proteins, band 3 and glycophorin C. The spectrin is polymeric, constructed from units which are tetrameric associations of two heterodimers in head-to-head association. Each junction of the protein network is a protein complex formed by the band 4.1-assisted binding of approximately six spectrin tetramers to one actin, with one band 4.1 molecule for every spectrin tetramer at the spectrin binding site on actin (Bennett and Baines 2001). Each junction complex is coupled to the plasma membrane by the binding of each of its approximately six band 4.1 molecules to a glycophorin C molecule of the plasma membrane (Workman and Low 1998). In addition, one ankyrin molecule binds to a site near the midpoint of each spectrin tetramer of the membrane skeleton and to a band 3 tetramer of the plasma membrane (Van Dort et al. 1998). Of these two protein linkages, the latter is known qualitatively to be much stronger. A detailed review of the structural organization of the RBC membrane is given in Appendix A of Lim (2003).

From our point of view, the main role of the membrane skeleton is to provide an elastic net which inhibits strong local deformations of the plasma membrane. The main elastic components of the membrane skeleton are the spectrin polymers, which form the connections between the vertices of the triangular mesh. Each unit of the spectrin is an $(\alpha\beta)_2$ double-stranded tetramer with an overall contour length of 200 nm. The persistence length of these linear units is only 10–20 nm at physiological temperatures. Thus, they are rather flexible, a significant part of their elasticity is entropic, and they will tend at equilibrium to adopt a partially folded configuration. As a consequence, in the resting state of the membrane skeleton, the actual separation between the vertices of the net is about 76 nm and the offset from the plane of the plasma membrane is in the range of 30–50 nm (Boal 1994; Heinrich et al. 2001). The motion of membrane molecules with large cytoplasmic domains is known to be significantly inhibited, due at least in part to interference from the membrane skeleton.

2.2.3
More on the SDE Sequence of Cell-Shape Transformations

The standard terminology (Bessis 1972; Bessis 1973; Bessis 1974; Bessis 2000) for the shapes of the SDE sequence was introduced in Section 2.1.1 and Fig. 2.2. The major non-discocyte stages, stomatocyte (after Greek *stoma*, mouth) and echinocyte (Gr. *echinos*, sea urchin or hedgehog), are each divided into four subcategories as follows (arranged in order of increased movement away from the discocyte in the two opposite directions):

Stomatocyte I: A cup shape with a shallow circular invagination.

Stomatocyte II: A cup shape with a deeper invagination, still at least approximately circular.

Stomatocyte III: A cup shape with a deep invagination, often elongated into a mouth-like slit and sometime accompanied by other pit-like invaginations.

Sphero-stomatocyte (or Stomatocyte IV): A spherical shape with small interior buds still attached to the membrane.

Echinocyte I: A disc with several undulations around its rim.

Echinocyte II: A flattened elliptical (oblate) body with rounded spicules distributed more or less uniformly over its surface.

Echinocyte III: An ovoid or spherical body with sharper and more numerous (30–50) spicules distributed evenly over its surface.

Sphero-echinocyte (or Echinocyte IV): A sphere with small sharp projections still attached to its surface.

The sequence of shape transformations listed above and depicted in Figs. 2.2 and 2.3 (without stage IV) is commonly known in the hematology literature, for example (Lichtman et al. 2005), as the stomatocyte-discocyte-echinocyte (SDE) shape transformations. The observations leading to the pictorial classification above are based on the behavior of populations of cells. Not surprisingly, with the natural variation of cellular properties, there is a spread of cell morphologies across several stages in any particular population at a given strength of the inducing agent (Hochmuth and Mohandas 1972; Seigneuret and Devaux 1984; Ferrell et al. 1985; Reinhart and Chien 1986; Rasia and Bollini 1998; Gedde et al. 1999); however, all cells can be driven to the terminal, stage IV, shapes.

In the classical literature the trajectory through stages I-II-III are regarded as reversible, in the sense that removal of the inducing agent or addition of an "antagonist" agent at stage III can cycle the shapes back through II and I and so forth. We will find in Section 2.7.3 that this reversibility is probably to some extent imperfect, that is, that some shape "hysteresis" is expected. On the other hand, once stage IV (the "sphero-" stage) is reached, reversal is no longer possible. Indeed, careful observation shows that the overall area of the visible membrane (which remains constant during earlier-stage shape changes) decreases in stage IV. At the same time, small vesicles composed apparently of plasma-membrane material are observed in the cytosol of sphero-stomatocytes and in the extra-cellular fluid for sphero-echinocytes. It is inferred that these vesicles (which lack membrane skeleton) have budded off from the plasma-membrane. This budding process probably takes place without detachment of the cytoskeletal proteins that anchor the membrane skeleton to the plasma membrane, via direct flow of plasma-membrane material into the small regions between the anchoring proteins.[8] In any case, our calculations will not address the terminal, stage IV, shape classes. Rapid hemolysis (rupture of the RBC membrane) occurs on further forcing of the terminal stages.

In addition to shapes belonging to the dominant SDE sequence, other minor shape classes can occur naturally or by design. Three of these are of particular interest to us as a consequence of their being found in the catalogue of shapes predicted in this work. These three shapes, illustrated in Fig. 2.3 (A, B and C), are:

Non-axisymmetric discocyte: A discocytic shape but with an uneven rim thickness, retaining the up-down symmetry of the normal discocyte but having a single additional mirror plane instead of full axisymmetry, produced by treating an osmotically swollen, nearly spherical RBC with diamide and then osmotically shrinking it back to the normal volume (Fischer et al. 1981).

---

[8] An alternative mechanism would be a simple disjoining of the plasma membrane from the membrane skeleton via the pulling out of the cytoskeletal anchors due to the pressure Q that acts between the two parts of the composite membrane (Section 2.4.1). However, a crude estimate (Mukhopadhyay et al. 2002) suggests that the required pull-out pressure remains two orders of magnitude above the pressures which actually occur.

Triangular stomatocyte: A cup shape with a deep triangular invagination. A stomatocyte II sometimes transforms to this shape, instead of a stomatocyte III shape, when treated with a stomatocytic agent (Bessis 1972).

Knizocyte: A triconcave RBC (Bessis 1973; Bessis 1974) found predominantly in healthy newborn infants (Ruef and Linderkamp 1999) but also observed in certain hemolytic anemias, such as hereditary spherocytosis (Bessis 1972).

Normal red cells subject to appropriate forcing adopt the shapes listed above. There exist additional shape classes associated with erythrocytes which have significant structural abnormalities. These include elliptocytes (Liu et al. 1982) (biconcave discocytes deformed so they have an elliptical outline), which apparently occur when the membrane skeleton is weakened or absent as occurs in hereditary elliptocytosis or in the lab when red cells are subjected to urea treatment (Khairy et al. 2007). In addition, we mention reticulocytes (Section 2.1.2), in which the structure of the mature red cell has not yet fully developed; codocytes, in which the area of the cell is abnormally large relative to its volume; keratocytes and acanthocytes, in which the membrane skeleton is damaged or deformed; and sickle cells, in which the hemoglobin carried by the cell polymerizes, so that the cytosol is no longer fluid and the cell membrane is no longer the principal determinant of cell shape. While certainly interesting, these shapes will not be discussed further herein.

2.3
Membrane Energetics

Membrane mechanics is represented mathematically by giving the membrane (free) energy as a functional $F[S]$ of the membrane configuration S. Because the membrane thickness is small on the scale of the red cell (and, also, on the scale of typical radii of curvature of red-cell surface features), it will suffice to treat S as a strictly two-dimensional surface. In this section we introduce the various contributions to $F[S]$.

We will organize these contributions into three classes: contributions $F_{\text{con}}[S]$ associated with the constraints of fixed membrane area A and cell volume V; contributions $F_{\text{pm}}[S]$ associated with the bending resistance of the plasma membrane; and contributions $F_{\text{ms}}[S]$ associated with the stretch and shear rigidity of the membrane skeleton. These contributions are additive at the level of our description,

$$F[S_0; S] = F_{\text{con}}[S] + F_{\text{m}}[S_0; S], \tag{2.6}$$

where

$$F_{\text{m}}[S_0; S] = F_{\text{pm}}[S] + F_{\text{ms}}[S_0; S] \tag{2.7}$$

refers to the (free) energy of the membrane at fixed area and volume. Note that F_{ms}, but not the other contributions, depends on the notional undeformed or relaxed shape S_0 of the membrane skeleton, as will be discussed further below. This section presents a discussion of each of these contributions in turn.

For flaccid red cells the constraint term effectively sets the volume $V[S]$ and area $A[S]$ of the red cell to their normal resting values, V_0 and A_0. In this situation F_{con} is negligible and Eq. (2.6) reduces to Eq. (2.7). A term like $F_{\mathrm{con}}[S]$ but with different coefficients will reappear in Section 2.5 as a convenient computational device.

Once the shape-energy functional is known, the problem of finding equilibrium shapes reduces to solving the variational problem $\delta F[S] = 0$ (or $\delta F_{\mathrm{m}}[S] = 0$), which is, of course, equivalent to mechanical equilibrium under Newton's laws. In particular, mechanically stable shapes are (local or global) minima of $F[S]$. To find such energy-minimizing shapes we have always two choices. We can solve Newton's equations or we can simply search $F[S]$ directly for minima. In Section 2.4 and Appendix D we will derive and discuss the form of the equations of Newtonian membrane statics, starting with F_{pm} and then adding in F_{ms}. While these equations have conceptual and historical interest, they are not – except in the very simplest cases – analytically soluble. Indeed, they have so far proved numerically tractable only for axisymmetric geometries. Thus, we will go on in Section 2.5 to the explanation of how to implement numerically the direct search for minima. It is this method which will form the basis of the cell-shape calculations reported in Section 2.6.

2.3.1
Energies of Constraint

The constraint energy has two terms, one associated with the cell area and the other with the cell volume,

$$F_{\mathrm{con}}[S] = F_A[S] + F_V[S]. \tag{2.8}$$

F_A expresses the elastic energy of area dilation/compression when the actual cell area $A[S]$ is forced to differ from its relaxed value A_0 (which for us will usually be $A_0 = A_{RBC}$). Of course, dilating the area causes stresses in both the plasma membrane and the membrane skeleton; however, as noted in Section 2.1.3, the elastic moduli for the skeleton are much smaller than those of the plasma membrane. Thus, we approximate (Evans and Skalak 1980; Seifert 1997),

$$F_A[S] \approx \frac{K_A \left(A[S] - A_0\right)^2}{2 A_0}, \tag{2.9}$$

valid for small deviations of A from A_0, where K_A is the stretch or area modulus of the plasma membrane, whose measured value is given in Table 2.2.

F_V expresses the osmotic free energy caused by changing the volume of the cell from its natural value V_0, at which it is at osmotic equilibrium, to some other value $V[S]$, at which it is not. Suppose that the initial equilibrium solute concentration inside the cell is $c_0 = n_0/V_0$, where n_0 is the number (in osmoles) of osmotically active molecules trapped within the plasma membrane. When the cell has the larger volume V, the solute concentration is reduced to $c = n_0/V$ and the van't Hoff relation requires an osmotic pressure difference across the membrane given by (Seifert 1997)

$$\Delta P(V) = RT(c_0 - c) = RTn_0 \left(\frac{1}{V_0} - \frac{1}{V}\right), \quad (2.10)$$

where $R = 8.314\,\text{Jmol}^{-1}\text{K}^{-1}$ is the universal gas constant and T is the absolute temperature (measured in Kelvin). Integration of $\Delta P(V)$ with respect to volume from V_0 to V gives the work done in the expansion and, therefore, the osmotic energy stored,

$$F_V = \int_{V_0}^{V} dV\, \Delta P(V) = RT\left[c_0(V - V_0) - n_0 \ln\left(\frac{V}{V_0}\right)\right], \quad (2.11)$$

which becomes

$$F_V[S] \approx \frac{K_V(V[S] - V_0)^2}{2V_0} \quad (2.12)$$

in the limit of small deviations of V from V_0. $K_V \equiv RTc_0$ is called the "osmotic modulus." At physiological osmolarity and $T = 300\,\text{K}$, $K_V = 7.23 \times 10^5\,\text{J/m}^3$, as given in Table 2.2.

For flaccid cells, the energy scale is set by the bending modulus κ_b. Thus, we expect $F_A \sim F_V \sim \kappa_b$, which means that $\Delta A/A_0 \sim \Delta V/V_0 \sim 5 \times 10^{-5}$. For this reason, there is a negligible error in assuming – as we shall – that the area and volume are strictly fixed for flaccid cells. In analytic work for flaccid cells, we will enforce the constraint on red-cell area and volume strictly (see Section 2.4). On the other hand, in the numerical Monte Carlo simulations of Section 2.5, it is more convenient to allow the area and volume to vary subject to constraints of the form of Eqs. (2.9) and (2.12), which serve to make variations of area and volume about A_0 and V_0 negligibly small. In this context, there is no particular reason to use the physical values of the moduli K_A and K_V. Instead, we have used the weaker computational moduli K_A^* and K_V^* given in Table 2.2, which serve to set $A \approx A_0$ and $V \approx V_0$ to within tolerances of 0.02%, which is sufficient for our purposes and computationally more efficient than the harder constraints.

2.3.2
Bending Energy of the Plasma Membrane

We model the plasma membrane as an isotropic fluid bilayer in which each leaflet has a uniform composition. Uniformity is, of course, an approximation. We will comment further on this approximation in Section 2.8.6. This model without the cytoskeletal contribution has been used extensively to describe the properties of lipid-bilayer vesicles. The reader is directed to earlier literature (Miao 1992; Miao et al. 1994; Wortis and Evans 1997), and particularly the excellent review by Seifert (1997). In what follows we give a brief pedagogical presentation of this material.

The free-energy functional $F_{\text{pm}}[S]$ describes the bending resistance of the plasma membrane and consists of three terms,

$$F_{\text{pm}}[S] = F_{\text{sc}}[S] + F_{\text{g}}[S] + F_{\text{ad}}[S], \tag{2.13}$$

where S is the two-dimensional mathematical surface representing the closed bilayer. The forms of these three terms are given by

$$F_{\text{sc}}[S] = \frac{\kappa_{\text{b}}}{2} \oint_S dA \left[2H(\mathbf{r}) - C_0\right]^2, \tag{2.14}$$

$$F_{\text{g}}[S] = \kappa_{\text{g}} \oint_S dA \, K(\mathbf{r}), \tag{2.15}$$

$$F_{\text{ad}}[S] = \frac{\pi \overline{\kappa}}{2 D_0^2 A_0} \left(\Delta A[S] - \Delta A_0\right)^2, \tag{2.16}$$

which we shall refer to as the spontaneous-curvature (sc) or "Helfrich" term, the Gaussian-curvature (g) term and the area-difference (ad) term, respectively. Collectively these terms constitute what is called the area-difference elasticity (ADE) model. In the succeeding paragraphs we will explain the first two terms and then the third.

In the first two terms, $F_{\text{sc}} + F_{\text{g}}$, the integrals are surface integrals over S. The quantities $H(\mathbf{r})$ and $K(\mathbf{r})$ are, respectively, the mean and Gaussian curvatures of S at the point $\mathbf{r}$. Thus, if the principal radii of curvature of S at $\mathbf{r}$ are denoted $R_i(\mathbf{r}), i = 1, 2$ (so the principal curvatures are $C_i(\mathbf{r}) = 1/R_i(\mathbf{r})$), then

$$H(\mathbf{r}) = \frac{1}{2}\left(C_1(\mathbf{r}) + C_2(\mathbf{r})\right) = \frac{1}{2}\text{tr}\,\mathbf{C} \tag{2.17}$$

$$K(\mathbf{r}) = C_1 C_2 = \det \mathbf{C}, \tag{2.18}$$

where $\mathbf{C}$ is the curvature tensor (see Appendix C). The sign convention is chosen so that the C_i's are positive where the shape is convex (outward). κ_{b} and κ_{g} are the bending modulus (see Table 2.2) and the Gaussian modulus, respectively. C_0 is a material parameter called the spontaneous curvature (see below). Note that positive C_0 favors convex shapes.

The functional forms of F_{sc} and F_g follow from simple symmetry principles by a kind of Landau argument. We assume that the bending energy depends locally on S, so

$$F_{sc}[S] + F_g[S] = \oint_S dA\, f_b(\mathbf{r}), \tag{2.19}$$

where the bending (free) energy per unit area, f_b, depends on the local shape of S in the vicinity of $\mathbf{r}$ in a way which respects Euclidean invariance and the local in-plane isotropy of the fluid membrane. The upshot of this argument, which is presented in full in Appendix B, is that, through terms of order $(D/R)^2$,

$$f_b(\mathbf{r}) = \kappa_0 + \kappa_1(C_1 + C_2) + \kappa_3(C_1^2 + C_2^2) + \kappa_5 C_1 C_2, \tag{2.20}$$

where $C_{1,2}$ are the local principal curvatures of S in the vicinity of $\mathbf{r}$ and the coupling constants κ_i are material moduli independent of $\mathbf{r}$ (assuming that the membrane composition is uniform over S). Substitution of Eq. (2.20) into Eq. (2.19) gives an expression which is equivalent to $F_{sc} + F_g$ provided that the three material moduli $\kappa_1, \kappa_3,$ and κ_5 are related to the three parameters $\kappa_b, \kappa_g,$ and C_0 of Eqs. (2.14) and (2.15) according to $\kappa_b = 3\kappa_3$, $\kappa_b C_0 = -\kappa_1$ and $\kappa_g = \kappa_5 - 2\kappa_3$. Note that the term in κ_0 is shape-independent and, therefore, arbitrary for purposes of shape determination.

The Gaussian term F_g, Eq. (2.15), does not contribute to the shape problem and is often omitted in forming the sum F_{pm}, Eq. (2.13). The reason for this is a mathematical property called the Gauss–Bonnet theorem (Millman and Parker 1977), which guarantees that, for any smooth S, the surface integral $\oint dA\, K(\mathbf{r})$ of the Gaussian curvature is a topological invariant. For example, $\oint dA\, K(\mathbf{r}) = 4\pi$ for any surface with the topology of a sphere. It follows that this term can only distinguish between shapes of different topology but plays no role in selecting among shapes of the same topology.

We now address the third term F_{ad} of Eq. (2.13). This term arises because – as emphasized in Section 2.2 – the two leaflets of the plasma membrane do not interchange material readily and, therefore, may in principle have slightly different relaxed areas (due, for example, to a few additional molecules in one leaflet compared to the other or to a difference between leaflets in the average relaxed area per molecule). This relaxed area difference, denoted $\Delta A_0 = A_{out} - A_{in}$, was already introduced in Section 2.1.7 in connection with the bilayer–couple hypothesis, and we are now in a position to deal with it in a careful manner. If the two leaflets were maintained flat and with their edges fixed together, then for $\Delta A_0 > 0$ the material in the outer leaflet would have to be compressed and/or the material in the inner leaflet would have to be expanded.[9] This would cause a strain variation

---

9) A heated bimetallic strip provides a one-dimensional analog.

across the thickness of the membrane and would give rise to a bending moment or "couple" tending to make the membrane become convex outwards. The situation is similar for a closed membrane, which is necessarily curved. In this case, there is a (small) area difference,

$$\Delta A[S] \equiv 2D_0 \oint dA\, H(\mathbf{r}), \tag{2.21}$$

between the two leaves of the bilayer, where $D_0 \approx D/2$ is the distance – assumed uniform over the membrane – between the midpoints of the two leaflets.[10] This well-known relation is rederived in Appendix C. The generalization of the flat-membrane discussion above is the statement that a bending moment is produced whenever $(\Delta A[S] - \Delta A_0) \neq 0$, and the dimensionless area strain of the resulting mismatch is $(\Delta A[S] - \Delta A_0)/A[S]$. The corresponding modulus is essentially the area modulus K_A,[11] so the stored strain energy per unit area becomes $\sim K_A[(\Delta A[S] - \Delta A_0)/A[S]]^2$. Summing this over the entire membrane gives the area-difference strain energy,

$$F_{\text{ad}}[S] \sim A[S] \times \frac{K_A D_0^2}{D_0^2} \times \left(\frac{\Delta A[S] - \Delta A_0}{A[S]}\right)^2,$$

which is equivalent to Eq. (2.16) for $A[S] = A_0$. Note that $\overline{\kappa} \sim D^2 K_A \sim \kappa_\text{b}$, so $\alpha_\text{b} \equiv \overline{\kappa}/\kappa_\text{b}$ is expected to be of order unity (see Table 2.2 and Appendix A). Because of the forms of Eqs. (2.16) and (2.21), F_{ad} appears nonlocal and could not have been picked up by the Landau argument above. $\overline{\kappa}$ is called the area-difference modulus or, sometimes, the nonlocal bending modulus. The extra factor π in the numerator is purely conventional. A more detailed form of this discussion is given in Miao et al. (1994).

The values that we have used in calculation for the elastic moduli are $\kappa_\text{b} = 2.0 \times 10^{-19}$ J and $\overline{\kappa} = 2\kappa_\text{b}/\pi$, as shown in Table 2.2 and discussed further in Appendix A.

The reader will have noticed that positive spontaneous curvature C_0 in F_{sc} and positive area difference ΔA_0 in F_{ad} have similar effects in that they both promote outward convexity of the membrane. Indeed, it is convenient to combine Eqs. (2.14) and (2.16) to rewrite Eq. (2.13) in the form,

---

10) Technically, D_0 is the distance between the so-called "neutral surfaces" of the two leaflets (Evans and Skalak 1980), the point of the leaflet profile about which the bending moment vanishes. This point is not necessarily at the leaflet midpoint. However, the leaflet stress profile has not to our knowledge been reliably measured, so we adopt $D_0 \approx D/2$ as a working hypothesis. Since D_0 only appears in our calculations via the effective spontaneous curvature, Eq. (2.23), any change in its value serves only to modify the relation between the physical quantity ΔA_0 and the computational quantity $\overline{m}_0$ (see more at Eq. (2.101) and Section 2.8.5).

11) There are additional numerical factors, since K_A refers to the entire membrane while the energy here refers to the stretching/compression of individual leaflets (Miao et al. 1994).

$$F_{\text{pm}}[S] = \frac{\kappa_{\text{b}}}{2} \oint_S \mathrm{d}A\, [2H(\mathbf{r}) - \overline{C}_0]^2 + \frac{\pi \overline{\kappa}}{2 A_0} \left(\oint_S \mathrm{d}A\, 2H \right)^2 + \text{constant}, \quad (2.22)$$

where

$$\overline{C}_0 = C_0 + \frac{\pi \alpha_{\text{b}}}{D_0} \frac{\Delta A_0}{A_0} \quad (2.23)$$

acts as an effective spontaneous curvature and the constant term is shape-independent. In arriving at Eq. (2.22), the Gaussian term has been dropped and the area is assumed fixed at A_0, as is appropriate for flaccid vesicles. Equation (2.23) shows how C_0 and ΔA_0 combine to make up the single control parameter $\overline{C}_0$ introduced in Section 2.1.5.[12] It may be useful to distinguish the somewhat different origins of these two similar effects. C_0 is best thought of as arising from the shapes of individual molecules in the leaflets of the bilayer. Positive C_0 arises from a preponderance of large headgroups and/or small tailgroups in the outer leaflet or, correspondingly, a preponderance of large tailgroups and/or small headgroups in the inner leaflet. Both of these effects produce bending moments for the individual leaflets. If the leaflets are symmetrical, then the overall effect cancels for the bilayer and $C_0 = 0$. But, the plasma membrane is not symmetrical (Section 2.2.1), so there is no reason for such a cancellation to occur. By contrast, the area-difference effect occurs even when molecular-shape effects are completely absent, whenever more (or larger) molecules are packed into one leaflet relative to the other. Helfrich (1973) was the first to postulate the spontaneous-curvature mechanism. Helfrich (1974) and Evans (Evans 1974; Evans 1980) later identified the area-difference effect. It is now recognized that these effects are both present and generically comparable in magnitude.[13] Neither C_0 nor ΔA_0 can at this point be measured directly. The fact that they enter the shape problem together in $\overline{C}_0$ is very convenient, since there is only one unknown parameter, which must be inferred indirectly from observation.

2.3.3
Elastic Energy of the Membrane Skeleton

Following Evans and Skalak (1980), we model the membrane skeleton as a two-dimensional continuous isotropic hyperelastic material (Ogden 1984;

---

12) The two input variables C_0 and ΔA_0 appear *only* in the combination $\overline{C}_0$ and cannot, therefore, be deduced independently from experiment. Of course, our choice to think of the combined variable as an effective spontaneous curvature is arbitrary. We could equally well have defined as a control parameter the effective area difference, $\overline{\Delta A_0} = \Delta A_0 + \frac{D_0 A_0}{\pi \alpha_b} C_0$.

13) The limiting case $\overline{\kappa} \to \infty$, in which the area difference ΔA is locked to ΔA_0, is sometimes called the bilayer–couple or ΔA model. (Svetina et al. 1982; Svetina and Žekš 1983; Svetina et al. 1985; Svetina and Žekš 1985; Svetina and Žekš 1989)

Mase and Mase 1999; Başar and Weichert 2000; Holzapfel 2000) confined to the surface S. Technically, "hyperelastic" means that the elasticity is assumed non-dissipative[14]; however, we intend in addition to emphasize that we will be dealing with large strains, well beyond the realm of linear elasticity, with dimensionless strains (see Section 2.7.4) which may be of order unity. It is an approximation to refer the cytoskeletal stresses to S, since the membrane skeleton is typically offset from the plasma membrane by distances $D_{ms} \sim 50$ nm (Sections 2.1.2 and 2.2.2). Thus, we are ignoring corrections of relative order D_{ms}/R, where R is a typical radius of curvature of S. For smooth shapes, $D_{ms}/R \sim D_{ms}/R_A \sim 10^{-2}$, so these effects are small. We also ignore cytoskeletal bending moments. As we have seen in Section 2.3.2, any bending modulus due to the cytoskeleton should scale as $D_{ms}^2 K_\alpha$, where K_α is the cytoskeletal stretching modulus. Thus, we expect $\kappa_b^{ms}/\kappa_b \sim (D_{ms}/D)^2 (K_\alpha/K_A) \sim 10^{-3}$ (see Table 2.2), due to the weakness of K_α relative to K_A.

In this picture then, the membrane skeleton is a two-dimensional elastic continuum with spherical topology, which we visualize as starting from some initial shape S_0 at which it is uniform, isotropic and unstressed. We shall refer to S_0 as the "reference shape" of the membrane skeleton. A configuration of the membrane skeleton is a continuous one-to-one mapping of each point $\mathbf{R}_0$ of S_0 to a point $\mathbf{R}$ of S. There are, of course, many such mappings, since the material points of S_0 can be moved around on the mathematical surface S. Any such mapping produces a strain field over the now-deformed cytoskeletal material. This strain field can be mapped back to S_0. The integrated elastic energy of this strain field constitutes the cytoskeletal free energy F_{ms}.

In the mapping of S_0 to S, the unstressed two-dimensional neighborhood of each point $\mathbf{R}_0$ of S_0 is mapped to a (generally) strained two-dimensional neighborhood of the corresponding point $\mathbf{R}$ of S. This mapping involves Euclidean operations (translation and rotation), which have no effect on the elastic energy, plus a locally linear deformation. Because the local strain is linear, it transforms any infinitesimal circular domain of S_0 into what is generally an infinitesimal elliptical domain on S. The ratios by which the two elliptical axes on S have been stretched (or compressed) relative to the original circular domain on S_0 define the so-called principal extension ratios λ_1 and λ_2 associated locally with the mapping from S_0 to S. The principal extension ratios are in general dependent on the location $\mathbf{R}_0$.

The dependence of the local elastic energy density on these ratios is limited by two conditions. First, the energy density is minimum when there is no deformation and, second, deformations which are purely rotational are isometric and cannot change the energy. It follows from simple symmetry

14) Some authors use the term "Green elastic" to describe these materials.

arguments (see Appendix B) that the local elastic energy density can only depend on two "strain invariants." These invariants are conventionally chosen to be the so-called area and shear strains,

$$\alpha = \lambda_1 \lambda_2 - 1 \tag{2.24}$$

and

$$\beta = \frac{1}{2}\left[\frac{\lambda_1}{\lambda_2} + \frac{\lambda_2}{\lambda_1} - 2\right], \tag{2.25}$$

respectively. α measures the fractional area change as a local neighborhood on S_0 is mapped to S; β is a measure of shear strain. Both vanish when $\lambda_1 = \lambda_2 = 1$. They obey $\alpha > -1$ and $\beta \geq 0$. In analogy with Eq. (2.19), the general form of the membrane-skeleton free energy is assumed to be the integral of a local strain-energy density f_{ms},

$$F_{\mathrm{ms}}[S_0; S] = \oint_{S_0} dA_0 \, f_{\mathrm{ms}}(\alpha(\mathbf{R}_0), \beta(\mathbf{R}_0)), \tag{2.26}$$

where the integral is over the unstressed skeleton S_0 and the functional notation $[S_0; S]$ is intended to indicate that the energy depends on the mapping which distributes the material of S_0 onto S. By hypothesis, the unstressed (reference) state has the minimum energy density, which we may choose to vanish, so $f_{\mathrm{ms}}(\alpha, \beta) \geq f_{\mathrm{ms}}(0, 0) = 0$.

It remains only to specify the functional form of the local elastic energy density f_{ms}. At weak deformation the leading terms available are (Evans and Skalak 1980)

$$f_{\mathrm{ms}}(\alpha, \beta) \approx \frac{K_\alpha \alpha^2}{2} + \mu \beta, \quad \text{for small strains,} \tag{2.27}$$

which defines K_α and μ, the elastic moduli for cytoskeletal stretch/compression and shear, respectively. Experimental values for these linear moduli are given in Table 2.2 and discussed in Appendix A. However, there is no reason to expect that these "linear" expressions should remain valid when $|\lambda_{1,2} - 1|$ is *not* small with respect to unity, that is, when the strains α and β become of order unity, [15] as will turn out to occur for well-developed echinocytes and (to a lesser extent) stomatocytes (see Section 2.7.4). The origin of the cytoskeletal elasticity is the configurational entropy of the coiled polymeric units. Polymers become inextensible as they are pulled out towards the maximum length allowed by their covalent bonds. It follows that we expect a "hardening" of polymeric elasticity at moderate and large deformations. We represent this hardening empirically by adding higher terms to the linear form Eq. (2.27), so

---

15) In practice, nonlinear effects become noticeable when $|\lambda_{1,2} - 1| \sim 0.1$.

$$F_{\text{ms}}[S_0; S] = F_{\text{stretch}} + F_{\text{shear}} \qquad (2.28)$$

$$= \frac{K_\alpha}{2} \oint_{S_0} dA_0 \left(\alpha^2 + a_3 \alpha^3 + a_4 \alpha^4 \right)$$

$$+ \mu \oint_{S_0} dA_0 \left(\beta + b_1 \alpha\beta + b_2 \beta^2 \right), \qquad (2.29)$$

which is the expression for skeletal elasticity that we use in computation. The coefficients a_3, a_4, b_1 and b_2 are dimensionless higher-order nonlinear elastic moduli, whose computational values are given in Table 2.2. As described in detail in Appendix A, the origin of these values is a phenomenological fit which incorporates recent experiments at both low and high deformations. We emphasize that Eq. (2.28) has no fundamental significance but is just a way of extending the usual linear relation Eq. (2.27) to high deformations to accommodate the expected hardening effects in a flexible manner. What we will find (see Section 2.8.3) is that we need relatively weak linear moduli K_α and μ to reproduce the observed SDE sequence. However, without the nonlinear correction terms such weak linear moduli would not be consistent with high-deformation experiments.

Finally, we need to discuss the reference shape S_0 at which the membrane skeleton is assumed to be relaxed, that is, to have zero elastic energy. First, a technical point. In our model it costs no energy to bend the membrane skeleton. It follows that S_0 is not uniquely determined, since any shape can be creased arbitrarily without effect on its energy as long as local infinitesimal neighborhoods are not otherwise deformed: a simple example is local indentation of a spherical surface inside a circular perimeter, which – carried to the extreme – would reduce it to a doubled half sphere. This degeneracy may be resolved (although in practice we shall not always do so) by always choosing for the representative reference shape the "inflated" shape that one would obtain by very gently "blowing up" the shape to its maximal volume.

Note that the shape, area and volume of S_0 are generally unknown and may be different from those of S. We emphasize that there is no simple way in which the shape of S_0 can be determined experimentally by physically removing the membrane skeleton from the RBC membrane (Johnson et al. 1980; Lange et al. 1982; Vertessy and Steck 1989; Svoboda et al. 1992; Lenormand et al. 2001; Lenormand et al. 2003), since the properties of the membrane skeleton depend on its environment, that is, on the physicochemical properties of the closely associated cytoplasm and plasma membrane. For example, it was shown by Liu and Palek (1984) that intracellular hemoglobin promotes spectrin dimer-dimer association and spectrin tetramer stability. Therefore, the removal of hemoglobin will affect the geometry and area of S_0. For this reason, we are forced to treat S_0 as unknown and to try to determine it by fitting to experiment. To do this generally is

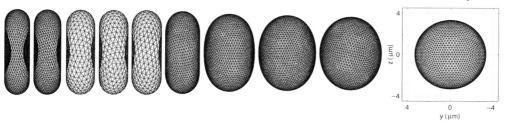

Fig. 2.7 Family of possible reference shapes S_0 assumed for the red-cell membrane skeleton. All these shapes have $A[S_0] = A_0$. The control parameter is the volume $V[S_0] \equiv V_{ms}$, as described in the text. Left to right: S_0 at $V_{ms} = 100, 110, 114, 116, 118, 130, 148, 152, 154$ and $155.8\ \mu m^3$.

computationally impractical; therefore, we will make simplifying assumptions. First, we assume that the area $A[S_0] = A_0$, that is, that the surface area of S_0 is that of the normal red cell. This may or may not be so;[16] however, as we shall see in Section 2.3.4 below, the principal effect of any difference in area between S and S_0 would be to modify the elastic constants K_α and μ. Since these are determined experimentally for intact red cells (see Appendix A), we may assume that such changes are already incorporated. Second, we restrict consideration to a particular plausible one-parameter class of possible reference shapes, shown in Fig. 2.7. The two limiting shapes are (at the left) a discocytic shape with volume V_0 and (at the right) a sphere of volume V_{A_0} (with $v = 0.642$ and $v = 1$, respectively, in reduced units). They correspond to the picture that the membrane skeleton has been moulded in some way to be stress free at the normal discocyte shape or, on the other hand, that it is a uniform sphere with no shape preference. Uniform tank-treading (Fischer and Schmidt-Schönbein 1977) in shear flow suggests the latter picture; however, recent "go-and-stop" experiments by Fischer (2004) provide evidence for some non-uniformity of the membrane. The intermediate shapes in Fig. 2.7 are a one-parameter interpolation between these two limits labeled by the volume $V_{ms} \equiv V[S_0]$ of the relaxed membrane skeleton. Specifically, this sequence is defined by minimizing $F_m[S]$ at fixed area and volumes $V_0 \leq V_{ms} \leq V_{A_0}$ with the following parameters, $\bar{\kappa} = C_0 = 0$, $K_\alpha = \kappa_b/\mu m^2$, $\mu = 0.2\,\kappa_b/\mu m^2$ and no nonlinear terms. This somewhat arbitrary choice was motivated by the fact that it is known from earlier work (Seifert et al. 1991) that such a sequence arises from the spontaneous curvature model Eq. (2.14) with $C_0 = 0$. The reason for adding a weak (c.f., Table 2.2) membrane-skeletal elasticity is numerical. As described in Section 2.5, we perform energy minimization computationally using a triangulated network. In the absence of skeletal elas-

---

16) Boal (1994) has suggested on the basis of a simulation model that the relaxed skeleton may be 10–20% smaller than the plasma membrane; on the other hand, Svoboda et al. (1992) find that isolated skeletons are expanded.

ticity, there is no driving force to keep the triangular elements fairly regular during Monte Carlo minimization.[17] As a result, the triangular elements have a tendency to become long and needle-like, so that they do not accurately represent a smooth surface. Introducing a weak F_{ms} is an effective way of keeping the shapes of the triangular elements regular, as Fig. 2.7 clearly shows.

2.3.4
Dimensionless Variables and Scaling

It is often convenient to work in dimensionless units. Such a representation is required for computational work. In this case the exercise of introducing such units will serve to highlight some important scaling properties of the membrane mechanics.

It is natural to scale all energies by the dominant energy κ_b and all lengths by the characteristic length R_A, thus defining, $h(\mathbf{r}) = R_A H(\mathbf{r})$, $c_0 = R_A C_0$, $a_0 = A_0/R_A^3 = 4\pi$, $da = dA/R_A^2$ and so forth.[18] In this rescaled notation, Eqs. (2.14) and (2.16) take the form,

$$\frac{1}{\kappa_b} F_{sc} = \frac{1}{2} \oint_S da \, [2h(\mathbf{r}) - c_0]^2 \qquad (2.30)$$

and

$$\frac{1}{\kappa_b} F_{ad} = \frac{\alpha_b}{8} (\Delta a[S] - \Delta a_0)^2, \qquad (2.31)$$

where $\Delta a[S] = \Delta A[S]/D_0 R_A$ and $\Delta a_0 = \Delta A_0/D_0 R_A$, so Eq. (2.22) becomes

$$\frac{1}{\kappa_b} F_{pm}[S] = \frac{1}{2} \oint_S da \, [2h(\mathbf{r}) - \bar{c}_0]^2 + \frac{\alpha_b}{2} \left(\oint_S da \, h(\mathbf{r}) \right)^2 + \text{constant}, \qquad (2.32)$$

where

$$\bar{c}_0 = c_0 + \frac{\alpha_b \Delta a_0}{4} \equiv \frac{\alpha_b}{2} \overline{m}_0 \qquad (2.33)$$

is the reduced effective spontaneous curvature and the quantity $\overline{m}_0$ is an equivalent reduced quantity commonly used in the literature.

The fact that the length scale R_A is absent from Eq. (2.32) means that the shape problem for the plasma membrane has a kind of scale invariance in that the energies of two bilayer vesicles (no membrane skeleton yet!) which differ only in scale are the same, provided that the corresponding values

---

17) This is related to the so-called "reparametrization invariance" of $F_{pm}[S]$, according to which a relabeling of the coordinates of S cannot affect the result.

18) From this point of view it would be natural to define $v \equiv V/R_A^3$, so $v_{A_0} = 4\pi/3$. Fortunately or not, the notation of Eq. (2.3) for the reduced volume is well-established, so we have, instead, $v = 3V/4\pi R_A^3$ with $v_{A_0} = 1$.

of C_0 and ΔA_0 are related in such a way that $\bar{c}_0$ is the same for the two vesicles. If this is true, then, for example, the minimum energy shape for the two vesicles will be the same, despite their difference in size. Because c_0 and Δa_0 occur in Eq. (2.32) only in the combination Eq. (2.33), the minimum-energy problem for a pure bilayer vesicle is fully characterized by only two parameters, the reduced spontaneous curvature $\bar{c}_0$ and the reduced volume v (plus, of course, the ratio α_b of the area-difference modulus $\bar{\kappa}$ to κ_b). The value of the bending modulus, κ_b, does not affect the minimum-energy shape. Of course, at nonzero temperature T, the ratio of the thermal energy $k_B T$ to κ_b does determine the size of thermal fluctuations.

Finally, we turn to the effect of the length scale on the membrane-skeleton energy $F_{\text{ms}}[S_0; S]$. Note that, if S and S_0 change scale together, then the local principal extension ratios λ_1 and λ_2 do not change. In this situation the strains α and β remain invariant, so

$$\frac{1}{\kappa_b} F_{\text{ms}}[S_0; S] = \frac{R_A^2 K_\alpha}{2\kappa_b} \oint_{S_0} da_0 (\alpha^2 + a_3 \alpha^3 + a_4 \alpha^4) \\ + \frac{R_A^2 \mu}{\kappa_b} \oint_{S_0} da_0 (\beta + b_1 \alpha \beta + b_2 \beta^2). \quad (2.34)$$

The ratio κ_b/μ defines the elastic length scale Λ_{el}, Eq. (2.5), so the dimensionless coefficient of the scaled shear term is just the ratio $(R_A/\Lambda_{el})^2$. (This ratio also multiplies the stretch term, since K_α and μ are comparable.) This shows that the cytoskeletal elastic contribution becomes increasingly important at large length scales, as discussed in Section 2.1.5. Equivalently, the elastic moduli must be rescaled to keep the dimensionless couplings constant to achieve scale invariance.

If, on the other hand, S is held fixed while S_0 is reduced in linear scale by a factor b, then the local principal extension ratios both increase by a factor of b. In this situation the shear strains β remain invariant but the new stretches become $\alpha' = b^2 \lambda_1 \lambda_2 - 1$, which does not involve a simple power of b. Simplification occurs when the area of S is held fixed, since $dA/dA_0 = \lambda_1 \lambda_2$, so

$$\oint_{S_0} dA_0 = A[S_0] \quad \text{and} \quad \oint_{S_0} dA_0 \lambda_1 \lambda_2 = A[S], \quad (2.35)$$

and only terms quadratic and higher in the product $\lambda_1 \lambda_2$ play a role in shape determination. It follows that, effectively, under $S_0 \to b^{-2} S_0$,

$$K_\alpha \oint_{S_0} dA_0 \, \alpha^2 + \mu \oint_{S_0} dA_0 \, \beta \longrightarrow b^2 K_\alpha \oint_{S_0} dA_0 \, \alpha^2 + \frac{\mu}{b^2} \oint_{S_0} dA_0 \, \beta. \quad (2.36)$$

Thus, at the harmonic (linear) level, decreasing the scale of S_0 by a linear factor b is equivalent to hardening the stretch modulus to $b^2 K_\alpha$ and softening the shear modulus to μ/b^2, as shown originally by Mukhopadhyay et al. (2002). The nonlinear elasticities do not scale so simply.

2.3.5
History: Other Red-Cell Models

Modern work on red-cell shapes dates to the work of Helfrich (1973), who focussed attention on the bending energy as the principal determinant of lipid-bilayer vesicle shapes. Indeed, the spontaneous curvature term, Eq. (2.14), is often called the "Helfrich model." The same energy functional was proposed by Canham (1970) only with the restriction $C_0 = 0$. It was recognized early-on (Deuling and Helfrich 1976) that, with appropriate choice of parameters, this model produces a variety of discocyte and stomatocyte shapes similar to red cells. The importance for lipid bilayers of the area-difference-elasticity term, Eq. (2.16), was first noted by Sheetz and Singer (1974), Helfrich (1974) and Evans (Evans 1974; Evans 1980) and was further developed by Svetina, Žekš and coworkers (Svetina et al. 1982; Svetina and Žekš 1983; Svetina et al. 1985; Svetina and Žekš 1985; Svetina and Žekš 1989) and many others (Seifert et al. 1991; Miao et al. 1994). The vesicle-shape problem has been recently reviewed in an excellent article by Seifert (1997). The importance of the cytoskeletal elasticity for the echinocytic red-cell shapes was first stressed by Waugh (Waugh 1996; Khodadad et al. 1996), Iglič (Iglič 1997; Iglič et al. 1998a; Iglič et al. 1998b) and others (Wortis 1998). Full calculations, based on both F_{pm} and F_{ms}, have only recently been reported in the literature (Lim et al. 2002; Mukhopadhyay et al. 2002; Lim 2003).

The earliest calculations of red-cell shapes based on membrane mechanics (Fung and Tong 1968; Zarda 1974; Zarda et al. 1977; Evans and Skalak 1980; Evans 1980; Pai and Weymann 1980; McMillan et al. 1986) incorrectly assumed that the membrane is a thin uniform isotropic elastic shell with no fluid component. The unstressed, preformed ("rubber-duck") shape was usually assumed to be either spherical or discocytic. These models give rise to a bending energy of the form,

$$\frac{\kappa_{\text{b}}}{2} \oint_{S_0} dA_0 \left[(C_1 - \widehat{C}_1)^2 + 2\nu(C_1 - \widehat{C}_1)(C_2 - \widehat{C}_2) + (C_2 - \widehat{C}_2)^2 \right], \quad (2.37)$$

where ν is a material parameter and $\widehat{C}_1(\mathbf{R}_0)$ and $\widehat{C}_2(\mathbf{R}_0)$ are the principal curvatures at the point $\mathbf{R}_0$ of the unstressed shell S_0. The bending modulus is usually taken to be in the range of our κ_{b}. In addition to the bending contribution, the thin-shell models include an in-plane elastic component similar to Eq. (2.28) only with a stretch modulus on the scale of the plasma membrane area modulus K_A and a shear modulus on the much weaker scale of our μ.

Although they persist in the engineering and materials literature, the thin-shell models of RBC shape are flawed in that they ignore the composite structure of the RBC membrane; nevertheless, for reasons that we now discuss, they do provide a reasonable fit to observed red-cell shapes in some cases.[19] Note that, when $\nu = 1$ and $\widehat{C}_1 = \widehat{C}_2$ and is independent of $\mathbf{R}_0$, as would be true for a spherical S_0, then Eq. (2.37) reduces to Eq. (2.14) and the thin-shell models are equivalent to $F_{sc} + F_{ms}$, that is, to our membrane model Eq. (2.7), only without the ADE term, Eq. (2.16), and with an anomalously large K_α. In this situation the large elastic stretch modulus correctly and strongly inhibits significant overall area changes of the red-cell membrane; however, it also inhibits local shape changes such as spicule formation, which are low-energy deformations in F_{sc} and only weakly discouraged by our elastic energy F_{ms}. The upshot is that these models often give qualitatively reasonable shapes for deformations like swelling which do not involve significant local cytoskeletal stretching relative to the normal discocyte. However, the lack of an effective spontaneous curvature $\overline{C}_0$ (that is, the lack of the ΔA_0 parameter of the ADE term) means that the basic driving force for the SDE shape sequence is missing. The failure to distinguish between the weak cytoskeletal elasticity and the strong fixed-area constraint of the plasma membrane means that these models cannot correctly describe strongly deformed shapes.

Physically the problem with the thin-shell models is that they do not properly represent the composite nature of the RBC membrane. The plasma membrane is locally incompressible (large K_A) but has no resistance to static shear deformations; on the other hand, the membrane skeleton resists shear but is relatively easily compressible ($0 < K_\alpha \sim \mu \ll K_A$). Of course, when the membrane skeleton is compressed or expanded locally, the proteins which anchor the skeleton must move relative to the incompressible 2D lipid-bilayer fluid with viscous drag and consequent dissipation. In the absence of local skeletal compressibility, ($K_\alpha = \infty$), the areal density of the skeleton remains constant, so that relative motion of the (incompressible) 2D fluid and the skeleton, although allowed, is not required. There is an interesting formal connection between the thin-shell model and the composite model when $K_A = K_\alpha = \infty$ (*not* the case for the red cell). In this situation, the ADE term, Eq. (2.16), vanishes, so both models are described by a bending elastic energy of the form of F_{sc}, Eq. (2.14), and a shear elasticity of the form of Eq. (2.27) (or Eq. (2.28)) with the area strain $\alpha = 0$ everywhere. The only difference between the two models is then in the spontaneous curvature C_0, which is in general spatially dependent ($C_0(\mathbf{r})$) for the thin-shell model but not for the spontaneous curvature model. Thus, for the special case of a strictly spherical S_0, the two models become indistinguishable in their static properties (although not, of course, for dynamics).

In the remainder of this section we discuss models of the composite RBC membrane and corresponding shape calculations. Generally speaking, these

---

[19] An example is the so-called "sphering" of the normal discocyte as it swells osmotically towards lysis (Zarda et al. 1977).

models are of two types. Some authors have represented the cytoskeleton at a coarse-grained level by a network of tethers or springs anchored to the plasma membrane at the junctions. For most of these "polymeric" models (Leibler and Maggs 1990; Discher et al. 1998; Li et al. 2005), the junction points of the network function computationally as a triangulation of the bilayer surface. An exception is the work of Boal et al. (1992) who used a finer triangulation for the fluid surface and a coarser one for the membrane skeleton. In this approach, the properties of the polymeric elements (and the temperature T) determine the effective elastic constants of the membrane skeleton. Thus, K_α and μ are automatically comparable and, when this approach is carried through carefully, the relation $K_\alpha \approx 2\mu$ emerges (Hansen et al. 1996; Boal 2002) and the elasticity hardens automatically at large deformations. On the other hand, the necessity of handling the polymeric degrees of freedom (particularly, if the elasticity is generated entropically) makes the calculations cumbersome, and systematic RBC shape calculations have not yet been carried out. Of the two papers which do calculate shapes, Discher et al. (1998) focus exclusively on RBC micropipette aspiration and Li et al. (2005) study optical tweezer experiments. Neither group has included the ADE term. All four of the above groups use a calculational representation of the bending energy that is known to be at best approximate (Gompper and Kroll 1997).

Other authors (Evans and Skalak 1980; Evans 1980; Elgsaeter et al. 1986; Stokke et al. 1986a; Stokke et al. 1986b; Peterson 1992a; Peterson et al. 1992b; Waugh 1996; Iglič 1997; Iglič et al. 1998a; Iglič et al. 1998b; Mukhopadhyay et al. 2002; Lim et al. 2002; Lim 2003; Kuzman et al. 2004) have represented the cytoskeleton as a 2D elastic continuum described by an energy of the form Eq. (2.26). Of these, Evans and Skalak (1980), Evans (1980), Peterson (1992a), Peterson et al. (1992b), Waugh (1996), Iglič (1997), Iglič et al. (1998a) and Iglič et al. (1998b) use $K_\alpha = \infty$, effectively reducing the problem to a thin-shell model. The remaining authors use free energies that are equivalent to our membrane model, Eq. (2.7), including both ADE terms and soft cytoskeletal moduli. A variety of different forms for the local elastic energy density are used in these works. Most authors used the linear elasticity, Eq. (2.27) or equivalent, without specifically including hardening at high deformation. The exceptions are Elgsaeter et al. (1986), Stokke et al. (1986a), Stokke et al. (1986b), Lim et al. (2002) and Lim (2003), who use the hardening representation Eq. (2.28).

Work on the pure-lipid models (without cytoskeleton) identified discocyte and stomatocyte shapes. In contrast, work on the models that incorporate membrane-skeleton elasticity has naturally focussed on echinocytes and spicule shapes. Spicule shapes have been calculated in Stokke et al. (1986b), Waugh (1996), Iglič (1997), Iglič et al. (1998a), Iglič et al. (1998b), Mukhopadhyay et al. (2002), Lim et al. (2002), Lim (2003) and Kuzman et al. (2004); however, for computational reasons, most of these authors assumed some parametrized form for the spicule shape and used the energy minimization to choose parameter values. The only work which leaves the full shape to be determined by energy minimization is that of Mukhopadhyay et al. (2002), Lim et al. (2002) and Lim (2003).

In the context of models including cytoskeletal elasticity, there have been only two groups that have examined the full SDE red-cell sequence and the parameters controlling it. The more recent is Lim et al. (2002) and Lim (2003), work which is fully reported here. The earlier work is that of Elgsaeter et al. (1986), Stokke et al. (1986a) and Stokke et al. (1986b). The title of this series of two papers emphasizes the hypothesis that the RBC membrane skeleton may be a gel, which in retrospect is probably not correct. Furthermore, the driving force for shape change is incorrectly taken to be the "osmotic tension" of the spectrin gel, instead of the bilayer area difference. Nevertheless, the authors came up with a set of energy contributions that are close to ours, including the use of low values for the elastic cytoskeletal moduli. Although they were not able to calculate shapes due to the relatively crude computer resources of the day, they did adopt a set of parametrized shapes of stomatocyte, discocyte and echinocyte (spicule) types and they identified energy-minimizing shapes of these classes, thus building up the first systematic "erythrocyte cell shape class diagram," a remarkable achievement for that time.

2.4
Equations of Membrane Shape Mechanics

2.4.1
Introduction

Observable equilibrium RBC shapes are (aside from thermal fluctuations) local minima of the membrane (free) energy $F_m[S_0; S]$ at fixed area A_0 and volume V_0. As such, these shapes obey the equations of mechanical equilibrium which characterize the composite membrane. In this section we describe and summarize the equations of static membrane equilibrium. Full derivations are given in Appendix D. In principle these equations – subject to appropriate boundary conditions – can be solved to find any mechanical-equilibrium shape of the RBC. Such configurations include not only the local-minimum shapes but also a variety of unstable equilibrium shapes which are not generally observable in the laboratory. Thus, any solution of the mechanical equations must be tested to see whether it is a true energy minimum or a saddle point, before it is accepted as an observable shape. Up to now the approach to equilibrium shapes via integration of the Newtonian equations of mechanical equilibrium has only proved useful for axisymmetric shapes. The obstacle in the more general case is the numerical difficulty of implementing the condition that the computational surface should close smoothly. The numerical method described in Section 2.5 is based on direct minimization of $F_\mathrm{m}[S_0; S]$. It is not restricted to axisymmetric shapes and it handles closure naturally.

According to the model we have adopted in Section 2.3, the plasma membrane and the membrane skeleton can slide freely with respect to one another, subject only to a viscous friction for relative motion in the transverse direction. It follows that, at equilibrium, the force between them must be normal to the common membrane surface S. We designate this normal force $Q(s_1, s_2)$, where (s_1, s_2) are general curvilinear coordinates on S (see Appendix C). Q is the force per unit area which the plasma membrane exerts on the cytoskeleton, defined positive when it is directed outward from the interior of the cell. By Newton's third law there is a reaction force, $-Q$ (that is, a force which acts inwards when $Q > 0$) on the plasma membrane. In mechanical terms, Q is the only coupling between the two components of the RBC membrane, beyond the condition that they both are located on the common surface S.

In this context, it will simplify the discussion to break it into two parts corresponding to the composite nature of the RBC membrane. Section 2.4.2 describes the mechanics of an idealized plasma membrane in the absence of cytoskeleton. This system is characterized by the (free) energy functional Eq. (2.13) of an ideal lipid-bilayer vesicle. There is only a small difference in the equilibrium mechanical equations between a lipid-bilayer vesicle and the plasma membrane of the full RBC. For the vesicle, the normal force per unit area is the (uniform) pressure difference ΔP between the fluid environments inside and outside of the vesicle (we define $\Delta P > 0$ when the inside pressure is higher than the outside pressure). For the full RBC plasma membrane, the effect of the additional normal force due to the membrane skeleton is to replace this by the (generally nonuniform) effective pressure $\Delta P - Q(s_1, s_2)$.

Section 2.4.3 describes the mechanics of the membrane skeleton. The physics here is that of equilibrium two-dimensional (nonlinear) elastic theory. The complication is that the two-dimensional sheet is not flat but, rather, deformed to fit the curved metric of the surface S. The condition linking the membrane-skeleton mechanics with that of the plasma membrane is that the normal forces Q on the membrane skeleton and $-Q$ on the plasma membrane are an action-reaction pair in the sense of Newton's third law. Solution of the full, coupled problem is discussed briefly at the end of Section 2.4.3 and a simple axisymmetric example is provided in Appendix D.6.

Finally a comment on presentation. In treating mechanical equilibrium, there are always two equivalent complimentary approaches: the "Newtonian" approach which focusses on the balance of forces and the "energetic" approach which focusses on the stationary property of the energy. While we will use both approaches, we will tend to emphasize the former. The reason for this is historical and pedagogical. Much of the work in this field (but by no means all) has taken the energetic approach. The Newtonian approach provides good physical insight and has been somewhat neglected in the literature.

2.4.2
Mechanics of the Plasma Membrane

There are excellent reviews of the statics of fluid membranes with bending rigidity, including Evans and Skalak (1980), Yeung and Evans (1995), Seifert (1997), Powers et al. (2002), Capovilla and Guven (2002), Lomholt and Miao (2006), Fournier (2007) and Van Hemmen and Leibold (2007).

Fluid membrane mechanics takes as its starting point the energy functional $F_{\mathrm{pm}}[S]$ of the plasma membrane, Eq. (2.13). To incorporate the constraints of fixed area A_0 and volume V_0, it is convenient to introduce Lagrange multipliers Σ and ΔP and and to construct the variational functional,

$$F_{\mathrm{var}}[S] = F_{\mathrm{pm}}[S] + \Sigma A[S] - \Delta P V[S]. \tag{2.38}$$

Making F_{var} stationary with respect to unconstrained variations of S, that is, setting $\delta F_{\mathrm{var}} = 0$, now leads to shape equations parametrized by Σ and ΔP. By appropriately choosing the Lagrange multipliers, one sets the area and volume to A_0 and V_0, respectively. The parameter ΔP turns out to be the fluid-pressure difference across the membrane (as indicated by the notation); but, the parameter Σ is not, in general, the mechanical membrane tension, as we shall discuss further below.

A significant simplification is now possible because of the variational structure of Eq. (2.38). The entire S dependence of the area-difference term F_{ad}, Eq. (2.16), comes from the term $\Delta A[S]$, which in turn is related to the integral over the mean curvature by Eq. (2.21). Thus,

$$\delta F_{\mathrm{ad}}[S] = 2\kappa_{\mathrm{b}} \frac{\pi \alpha_{\mathrm{b}}}{D_0 A_0} (\Delta A[S] - \Delta A_0) \, \delta \oint_S dA \, H(\mathbf{r}). \tag{2.39}$$

On the other hand, the variation of the spontaneous-curvature term F_{sc} Eq. (2.14) takes the form,

$$\delta F_{\mathrm{sc}}[S] = -2\kappa_{\mathrm{b}} C_0 \, \delta \oint_S dA \, H(\mathbf{r}) + 2\kappa_{\mathrm{b}} \, \delta \oint_S dA \, H^2(\mathbf{r}), \tag{2.40}$$

which also contains a term proportional to the variation of the integrated mean curvature. It follows, therefore, that the entire effect of $F_{\mathrm{ad}}[S]$ on any variation of $F_{\text{pm}}[S]$ is effectively to replace C_0 in Eq. (2.40) by the combination,

$$C_0^{\text{eff}}[S] \equiv C_0 - \frac{\pi\alpha_{\text{b}}}{D_0 A_0}(\Delta A[S] - \Delta A_0) = \overline{C}_0 - \frac{\pi\alpha_{\text{b}}}{D_0}\frac{\Delta A[S]}{A_0}. \quad (2.41)$$

The equations of membrane mechanics arise from such variations. Furthermore, the variations of the Gaussian term F_{g} vanish, provided there is no change of topology. We conclude finally that the vesicle-shape mechanics of the full F_{pm} is equivalent to that of a pure spontaneous-curvature model (F_{sc} only), provided that C_0 is replaced by C_0^{eff} according to Eq. (2.41). Thus, in discussing fluid-membrane mechanics, there is no loss of generality in starting from F_{sc}, Eq. (2.14) (or, in case the Gaussian term comes in, from $(F_{\text{sc}} + F_{\text{g}})$).[20] In what follows we will take this approach. Furthermore, to keep the notation simple, we will write the spontaneous curvature parameter that appears in the mechanical equations as C_0, omitting the superscript and thereby suppressing the self-consistency.

2.4.2.1 Fluid Membrane Without Bending Rigidity

It is instructive to start the treatment of fluid-membrane mechanics with a simple example. Consider a bilayer–fluid vesicle with a relaxed membrane area A_0 surrounding a volume V_0 of incompressible aqueous solution. Assume that the membrane is completely characterized by a linear expansion modulus K_A, so that the energy of the membrane is

$$F[S] = \frac{K_A}{2}\left(\frac{A[S] - A_0}{A_0}\right)^2 A_0. \quad (2.42)$$

Equilibrium shapes correspond to minima of $F[S]$ with respect to variations of S at constant $V[S] = V_0$. There are two cases. For $V_0 < V_{A_0}$ (the "flaccid" vesicle of Section 2.1.3), there are many shapes with $A[S] = A_0$, that is, with zero membrane tension. All these shapes are (degenerate) energy minima, there is no unique ground-state shape, and there is no pressure difference across the membrane.[21]

On the other hand, for $V_0 > V_{A_0}$ (the "turgid" vesicle of Section 2.1.3), $A[S]$ must be greater than A_0, so the energy Eq. (2.42) cannot be zero. The equilibrium shape must minimize $A[S]$ at constant volume and is, therefore, spherical. The corresponding isotropic tension in the membrane at equilibrium is $\tau_0 \equiv K_A(A[S_{\text{min}}] - A_0)$. The same calculation can be carried out via

---

20) Of course, C_0^{eff} depends on S, making the full shape problem self-consistent from this perspective; however, this should not obscure the fact that C_0^{eff} appears parametrically in the mechanical equations.

21) This discussion ignores thermal fluctuations, which can generate both pressure and tension by forcing the system to sample configurations with $A[S] > A_0$. Such effects lead to $\left|\frac{A-A_0}{A_0}\right| \sim \left(\frac{k_bT}{K_A}\right)^{1/2} \sim 10^{-9}$ (for the typical parameters of Table 2.2) and are generally unimportant.

an unconstrained variation of S by constructing the variational functional (c.f., Eq. (2.38)),

$$F_{\text{var}}[S] = F[S] - \Delta P\, V[S]. \tag{2.43}$$

Varying with respect to S at constant $V[S]$ identifies $S_{\min}$ as a sphere of (unknown) radius R, so

$$F_{\text{var}}[S] \to F_{\text{var}}(R) = \frac{K_A}{2}\left(\frac{4\pi R^2 - A_0}{A_0}\right)^2 A_0 - \frac{4\pi}{3}\Delta P\, R^3. \tag{2.44}$$

Setting $dF/dR = 0$ evaluates the Lagrange multiplier,

$$\Delta P = \frac{2\tau_0}{R}, \tag{2.45}$$

which is just the usual soap-bubble equation.[22] Note that τ_0 depends on R, which in turn depends on V_0, so Eq. (2.45) gives the Lagrange parameter ΔP as a function of V_0. Finally, to identify ΔP with the mechanical pressure difference, we equate the axial forces on the spherical patch shown in Fig. 2.8. The upward force due to the pressure difference is $\pi r^2 \Delta P$; the downwards force due to the isotropic surface tension τ_0 is $2\pi r \sin\theta\, \tau_0$; $r = R\sin\theta$; and, Eq. (2.45) reemerges. Notice here that the nonzero tension and the nonzero pressure difference arise together as a result of the constraint of fixed volume and the area response of the stressed membrane.

A related but different situation arises for an isotropic fluid membrane of relaxed area A_0 spanning a hole in a partition between two compartments maintained at a fixed pressure difference ΔP (Fig. 2.9). If the aperture is circular, then it is not hard to show that the equilibrium membrane shape is again spherical and Eq. (2.45) holds. However, if the aperture is not circular, then the equilibrium membrane shape cannot be spherical. Suppose the membrane shape S is described as a vector function $\mathbf{R}(s^1, s^2)$ of general surface coordinates (s^1, s^2).[23] The membrane tension remains uniform,

$$\frac{\partial \tau_0}{\partial s^\alpha} = 0, \tag{2.46}$$

however, Eq. (2.45) generalizes to

$$\Delta P = \tau_0 \left(\frac{1}{R_1} + \frac{1}{R_2}\right) = \tau_0(C_1 + C_2) = 2\tau_0 H = \tau_0 \operatorname{tr} \mathbf{C}, \tag{2.47}$$

where R_i and C_i are the principal radii and principal curvatures, respectively, at the general point (s^1, s^2) of the membrane and $\mathbf{C}$ is the local 2×2 curvature tensor. Equation (2.47) may be derived as an exercise in variational

---

[22] Of course, for the soap bubble, τ_0 is fixed by the air-water interfacial tension and, thus, independent of R.

[23] The use of superscripts rather than subscripts here and in the following is motivated by the fact that the infinitesimal ds^α behaves under coordinate transformations in a contravariant manner (see Appendix C).

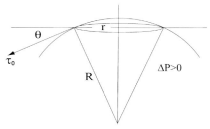

Fig. 2.8 Forces on patch with circular boundary. The upward force $\pi r^2 \Delta P$ due to the pressure difference balances the net downwards tension force applied tangentially along the perimeter of the patch.

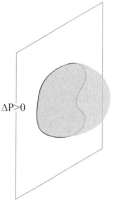

Fig. 2.9 Isotropic fluid membrane across non-circular aperture. The pressure on the left is higher than that on the right. Because the aperture perimeter is not circular, the membrane shape is not spherical. But, it does satisfy Eqs. (2.46) and (2.47).

calculus or, alternatively, by requiring that the net force on a small patch of membrane should vanish (see Eq. (2.56)). The curvature tensor $\mathbf{C}(s^1, s^2)$ is defined in terms of derivatives of $\mathbf{R}(s^1, s^2)$ (Appendix C), so Eq. (2.47) becomes a differential equation for $\mathbf{R}(s^1, s^2)$. The solution of Eqs. (2.46) and (2.47) subject to the appropriate boundary conditions determines both the equilibrium film shape S and the unknown membrane tension τ_0. For general $0 < K_A < \infty$, both S and τ_0 vary with the pressure difference ΔP, since they are related by $\tau_0 = K_A(A[S] - A_0)$. Note that the minimizing shape S is a surface of constant mean curvature equal to $\Delta P/2\tau_0$, which becomes a "minimal surface" ($H = 0$ everywhere) as $\Delta P \to 0$.

The relative areas of the hole (A_h) and the relaxed membrane (A_0) play an important role at low pressure differences. When $A_h > A_0$, the membrane must always be under tension to span the hole, so, as $\Delta P \to 0$, the equilibrium shape S becomes flat but the tension τ_0 remains positive. On the other hand, when $A_h < A_0$, the excess area means that, as $\Delta P \to 0$, τ_0

goes to zero and the minimum-energy shape is degenerate. The case of an effectively incompressible membrane is special (and relevant to the RBC). In this situation, $K_A \to \infty$ and $A[S] \to A_0$, so the product τ looks indeterminate. In this limit, the minimizing shape S becomes independent of ΔP for all $\Delta P > 0$. The ratio $\Delta P/\tau_0$ is fixed by geometry, that is, by the boundary conditions applied to Eq. (2.47). As a consequence, τ_0 scales linearly to zero with ΔP.[24] The limits $K_A \to \infty$ and $\Delta P \to 0$ do not commute.

We will see below how all this changes when the membrane acquires a bending rigidity.

2.4.2.2 General Equilibrium Conditions for Membranes with Internal Stresses

Before introducing bending rigidity, it will be useful to develop in a general way the conditions for mechanical equilibrium of membranes with internal stresses. These conditions derive from the laws of Newtonian statics, according to which an extended object can be in mechanical equilibrium if and only if the net force on it, $\mathbf{F}_{\text{net}}$, and the net torque on it, $\mathbf{N}_{\text{net}}$, both vanish. We assume for simplicity that there are only two kinds of forces which act on the membrane, external forces due to pressure differences across the membrane which act along the outward normal $\hat{\mathbf{n}}$ and internal forces related to the membrane stresses.

Consider a patch Σ of membrane described by the 3D vector function $\mathbf{R}(s^1, s^2)$ and surrounded by a boundary $\partial\Sigma$. The membrane-stress force on each infinitesimal element of $\partial\Sigma$ is proportional to its length dl and has in general the form $dl\, \mathbf{T}(\hat{\mathbf{p}})$, so $\mathbf{T}(\hat{\mathbf{p}})$ is a force per unit length. The argument here is entirely general; when necessary, we will distinguish membrane stresses due to the plasma membrane and the membrane stresses due to the membrane skeleton by writing $\mathbf{T}_{\text{pm}}(\hat{\mathbf{p}})$ and $\mathbf{T}_{\text{ms}}(\hat{\mathbf{p}})$, respectively, so that for a patch of the composite membrane,

$$\mathbf{T}(\hat{\mathbf{p}}) = \mathbf{T}_{\text{pm}}(\hat{\mathbf{p}}) + \mathbf{T}_{\text{ms}}(\hat{\mathbf{p}}), \tag{2.48}$$

and similarly for for other related quantities. In all these expressions, $\hat{\mathbf{p}}$ is the outwardly-directed in-plane unit normal to the boundary element $\partial\Sigma$ (see Fig. 2.10) and its appearance as a argument of $\mathbf{T}$ recognizes that the force per unit length will generally depend on the orientation of the boundary. Note that $\mathbf{T}(\hat{\mathbf{p}})$ is a 3D vector (written in bold face) and may have both in-plane and out-of-plane components. The net force on Σ consists, therefore, of a sum of two terms, which must vanish as a condition of mechanical equilibrium,

$$\mathbf{F}_{\text{net}} = \Delta P \int_\Sigma dA\, \hat{\mathbf{n}} + \int_{\partial\Sigma} dl\, \mathbf{T}(\hat{\mathbf{p}}) = \mathbf{0}, \tag{2.49}$$

with $dA = \sqrt{g}\, ds^1 ds^2$ and $dl^2 = g_{\alpha\beta} ds^\alpha ds^\beta$, where $g_{\alpha\beta}$ is the metric tensor and $g = \det g_{\alpha\beta}$, in the notation of Appendix C.

[24] All this can be worked out explicitly when the hole is circular.

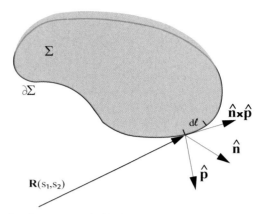

Fig. 2.10 Patch Σ of surface surrounded by boundary $\partial\Sigma$. At the boundary point $\mathbf{R}(s_1, s_2)$, $\hat{\mathbf{n}}$ is the unit outward normal to Σ and $\hat{\mathbf{p}}$ is the unit vector in the plane of Σ which points perpendicularly outward from $\partial\Sigma$. The boundary element dl is directed along $\hat{\mathbf{n}} \times \hat{\mathbf{p}}$.

In order to turn this global condition for translational equilibrium into a local one, we must rewrite the stress term making use of some properties of the boundary-stress forces. It follows from a simple argument (see Appendix D.1) that $\mathbf{T}(\hat{\mathbf{p}})$ can be decomposed as

$$\mathbf{T}(\hat{\mathbf{p}}) = p_\alpha \mathbf{T}^\alpha, \tag{2.50}$$

where $\hat{\mathbf{p}} = p^\alpha \mathbf{Y}_\alpha = p_\alpha g^{\alpha\beta} \mathbf{Y}_\beta$, in which $\mathbf{Y}_\alpha \equiv \partial \mathbf{R}/\partial s^\alpha$ are the tangent vectors to the surface (see Appendix C). The two components $\mathbf{T}^\alpha$, $\alpha = 1, 2$, are independent of $\hat{\mathbf{p}}$. We will refer to them collectively as the surface stress tensor. The lower indices on p_α and $\mathbf{Y}_\alpha$ indicate that these quantities transform under changes of the surface coordinates (s^1, s^2) as covariant first-rank tensors; the upper indices on p^α and $\mathbf{T}^\alpha$ indicate that these quantities transform as contravariant first-rank tensors. Contractions of upper and lower indices, such as occur in the expressions for the physical quantities $\hat{p}$ and $\mathbf{T}(\hat{p})$, are invariant under coordinate changes. Upper and lower indexed quantities are related by the metric tensor and its inverse,[25] $p^\alpha = g^{\alpha\beta} p_\beta$ with $g_{\alpha\beta} g^{\beta\gamma} = \delta^\gamma_\alpha$. The stress-tensor vector $\mathbf{T}^\alpha$ may be further decomposed into in-plane and out-of-plane components according to

$$\mathbf{T}^\alpha = T^{\alpha\beta} \mathbf{Y}_\beta + T^\alpha_\perp \hat{\mathbf{n}}, \tag{2.51}$$

in which the (physical) scalar quantities $T^{\alpha\beta}$ transform as a contravariant second-rank tensor under coordinate change. We are now in a position to transform the second term of Eq. (2.49) from a boundary integral to a surface integral,

25) We use $\delta^\beta_\alpha = \delta^\alpha_\beta = \delta_{\alpha\beta} = \delta^{\alpha\beta}$ with the convention of choosing the representation that makes the covariant notation fluent.

2.4 Equations of Membrane Shape Mechanics

$$\int_{\partial \Sigma} dl \, \mathbf{T}(\hat{\mathbf{p}}) = \int_{\partial \Sigma} dl \, p_\alpha \mathbf{T}^\alpha = \int_\Sigma dA \, D_\alpha \mathbf{T}^\alpha, \qquad (2.52)$$

where D_α is the covariant derivative, so $D_\alpha \mathbf{T}^\alpha \equiv \partial_\alpha \mathbf{T}^\alpha + \Gamma^\alpha_{\alpha\beta} \mathbf{T}^\beta$ is the curved-space version of the divergence. The equality of the second and third terms is the 2D curved-space version of Gauss's law (see Appendix C). By combining Eqs. (2.49) and (2.52) and shrinking the patch Σ to infinitesimal size, we find

$$\Delta P \hat{\mathbf{n}} + D_\alpha \mathbf{T}^\alpha = \mathbf{0}. \qquad (2.53)$$

Equation (2.53) is the general condition for local translational equilibrium with pressure and stress forces, as expressed in general surface coordinates, and will be the starting point of our discussion of the effects of membrane bending rigidity in the next section.

Before continuing it may be useful to see how this approach applies to the case of the fluid membrane without bending rigidity, as discussed in the previous section. In this case the stress acts perpendicularly outward on the boundary element, so

$$\mathbf{T}(\hat{p}) = \tau_0 \hat{p} = \tau_0 \mathbf{Y}^\alpha p_\alpha = \mathbf{T}^\alpha p_\alpha \qquad (2.54)$$

and

$$\mathbf{T}^\alpha = \tau_0 \mathbf{Y}^\alpha \qquad (2.55)$$

for membranes without bending rigidity. A brief calculation using results from Appendix C provides the evaluation, $D_\alpha(\tau_0 \mathbf{Y}^\alpha) = (\partial_\alpha \tau_0) \mathbf{Y}^\alpha - \tau_0 (\operatorname{tr} \mathbf{C}) \hat{\mathbf{n}}$, where $\mathbf{C}$ is the curvature tensor. Substituting this result into the general equilibrium condition, Eq. (2.53), gives

$$(\Delta P - \tau_0 \operatorname{tr} \mathbf{C}) \hat{\mathbf{n}} + (\partial_\alpha \tau_0) \mathbf{Y}^\alpha = \mathbf{0}, \qquad (2.56)$$

which is equivalent to the two conditions, Eq. (2.46) and (2.47), displayed above.

The second condition for mechanical equilibrium of the patch Σ is that the net torque $\mathbf{N}_{\text{net}}$ acting on it must vanish. This torque has contributions from the pressure and boundary forces. In addition, for membranes with bending rigidity, it may include an "intrinsic" torque $\mathbf{N}_{\text{int}}$, which we will discuss below. Thus, the torque condition takes the form,

$$\mathbf{N}_{\text{net}} = \Delta P \int_\Sigma dA \, \mathbf{R} \times \hat{\mathbf{n}} + \int_{\partial \Sigma} dl \, \mathbf{R} \times \mathbf{T}(\hat{\mathbf{p}}) + \mathbf{N}_{\text{int}} = \mathbf{0}, \qquad (2.57)$$

which is the analogue of Eq. (2.49) for the force equilibrium. Using Gauss's law simplifies the boundary integral,

$$\int_{\partial \Sigma} dl \, \mathbf{R} \times \mathbf{T}(\hat{\mathbf{p}}) = \int_{\partial \Sigma} dl \, \mathbf{R} \times \mathbf{T}^\alpha p_\alpha = \int_\Sigma dA \, D_\alpha (\mathbf{R} \times \mathbf{T}^\alpha)$$

$$= \int_\Sigma dA \, (\mathbf{Y}_\alpha \times \mathbf{T}^\alpha + \mathbf{R} \times D_\alpha \mathbf{T}^\alpha). \qquad (2.58)$$

When this is substituted into Eq. (2.57), the final term cancels against the pressure term by virtue of the force equilibrium condition, Eq. (2.49), and the result is

$$\mathbf{N}_{\text{net}} = \mathbf{N}_{\text{int}} + \int_\Sigma dA\, \mathbf{Y}_\alpha \times \mathbf{T}^\alpha$$

$$= \mathbf{N}_{\text{int}} + \int_\Sigma dA\, \mathbf{Y}_\alpha \times \left(T^{\alpha\beta}\mathbf{Y}_\beta + T^\alpha_\perp \hat{\mathbf{n}}\right) = \mathbf{0}, \quad (2.59)$$

which reduces finally to

$$\mathbf{N}_{\text{net}} = \mathbf{N}_{\text{int}} + \int_\Sigma dA\, \sqrt{g}(T^{12} - T^{21})\hat{\mathbf{n}} + \int_\Sigma dA\, T^\alpha_\perp \mathbf{Y}_\alpha \times \hat{\mathbf{n}} = 0. \quad (2.60)$$

It is often but not always true that $T^{\alpha\beta}$ is symmetric, so the second term vanishes (Lomholt and Miao 2006).

For the simple Helfrich model, Eq. (2.14), it will turn out that the form of the intrinsic torque is particularly simple. In this case the membrane rigidity produces an isotropic torque density of magnitude,

$$\mathcal{M} = \kappa_b(2H - C_0) = \kappa_b(C_1 + C_2 - C_0), \quad (2.61)$$

directed along the patch boundary in the sense $\hat{\mathbf{n}} \times \hat{\mathbf{p}}$, so

$$\mathbf{N}_{\text{int}} = \int_{\partial\Sigma} dl\, \mathcal{M}\hat{\mathbf{n}} \times \hat{\mathbf{p}} = \int_{\partial\Sigma} dl\, \left(\mathcal{M}\hat{\mathbf{n}} \times \mathbf{Y}^\alpha\right) p_\alpha$$

$$= \int_\Sigma dA\, D_\alpha\left(\mathcal{M}\hat{\mathbf{n}} \times \mathbf{Y}^\alpha\right)$$

$$= -\int_\Sigma dA\, g^{\alpha\beta}(\partial_\beta \mathcal{M})\mathbf{Y}_\alpha \times \hat{\mathbf{n}}, \quad (2.62)$$

where the last step requires some results from Appendix C. Combining this with Eq. (2.60) (and assuming $T_{\alpha\beta}$ to be symmetric) leads to the relation,

$$T^\alpha_\perp = g^{\alpha\beta}\partial_\beta \mathcal{M} = \partial^\beta \mathcal{M}, \quad (2.63)$$

which shows that the out-of-plane component of $\mathbf{T}^\alpha$, which vanishes in the absence of bending rigidity, is related to the local torque density when rigidity is present.

2.4.2.3 Fluid Membrane with Bending Rigidity

For lipid-bilayer membranes which are flaccid, that is, for $V[S] < V_{A_0}$ (Eq. (2.2)) or, generally, for membranes which are not constrained to exceed their relaxed area, the area compressibility does not generate membrane tension. In the absence of other effects, all flaccid configurations would be neu-

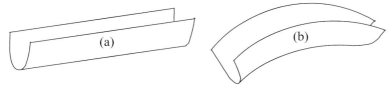

Fig. 2.11 Mechanical effect of Helfrich bending rigidity. When forced away from its preferred mean curvature C_0 by bending about one axis (a), the membrane responds by bending in the opposite sense about the perpendicular axis (b). In this illustration $C_0 = 0$, so the flat configuration is an energy minimum. When forced into the trough shape shown at the left, the membrane responds by adopting the saddle configuration shown at the right.

trally stable and there would be no unique energy-minimizing shapes. What breaks this degeneracy is the bending rigidity.[26]

It is not surprising that bending rigidity is associated with bending moments or torques in the membrane; however, it may be less clear how it generates tensions. Consider the form of the energy functional F_{sc}, Eq. (2.14). The energy density is minimum locally when $2H = C_0$. Suppose for simplicity that $C_0 = 0$, so that the flat configuration is an energy minimum. If we now bend the membrane, forcing it into a cylindrical trough of radius R along some chosen direction (Fig. 2.11), the membrane will try to respond by bending the trough axis about the perpendicular axis to achieve a complementary radius of curvature $-R$, so that the mean curvature again achieves $2H = 1/R - 1/R = 0$. If constraints inhibit the zero-energy state, then local torques will develop in the membrane. Equation (2.63) shows how these torques produce out-of-plane force densities on the scale $\mathcal{M}/R \sim \kappa_b/R^2$. In-plane effects are comparable. The upshot is that, in the presence of bending rigidity, the force density $\mathbf{T}(\hat{\mathbf{p}})$ develops a structure more complicated than that given by Eq. (2.54), with new terms both in-plane and out-of-plane. These new terms scale as κ_b/R^2. They are neither uniform nor isotropic. It is these terms which generate membrane tension even for flaccid vesicles.

The equilibrium mechanics of membranes governed by the energy Eq. (2.14) is summarized by giving the expression for the Helfrich stress tensor,[27]

$$\mathbf{T}^\alpha = \mathbf{T}^\alpha_{\text{pm}} = \left(\tau_0 + \frac{\kappa_b}{2}(2H - C_0)^2\right)\mathbf{Y}^\alpha - \mathcal{M}C^\alpha_\beta \mathbf{Y}^\beta + (\partial^\alpha \mathcal{M})\hat{\mathbf{n}} \quad (2.64)$$

or, equivalently,

---

26) In this discussion we will ignore thermal fluctuations, which typically produce effects of relative order $k_B T/\kappa_b \sim 1/50$. Such effects can, however, become important near the boundary between flaccid and turgid configurations, as has been examined in the vesicle context by Evans and Rawicz (1990).

27) This result requires generalization when external forces act on the membrane at positions which are not precisely at the center (neutral plane) of the membrane, as discussed by Lomholt and Miao (2006).

$$T_{\text{pm}}^{\alpha\beta} = \left(\tau_0 + \frac{\kappa_b}{2}(2H - C_0)^2\right)g^{\alpha\beta} - \mathcal{M}C^{\alpha\beta} \quad \text{and}$$

$$\left(T_{\text{pm}}\right)_\perp^\alpha = \partial^\alpha \mathcal{M}, \tag{2.65}$$

where $\mathcal{M}$ is given by Eq. (2.61). In the notation of Appendix C, the curvature tensor is $\mathbf{C} = C_\beta^\alpha = g^{\alpha\gamma}C_{\gamma\beta}$ and $2H = \text{tr } \mathbf{C}$. Note that the part of $\mathbf{T}^\alpha$ normal to the plane of the membrane obeys the general condition for torque equilibrium, Eq. (2.63). This result is derived in full in Appendix D.2 (Eq. (2.195)) and elsewhere (Evans and Skalak 1980; Yeung and Evans 1995; Capovilla and Guven 2002; Lomholt and Miao 2006). Equation (2.64) generalizes the simple soap-film stress tensor, Eq. (2.55). We will find in Appendix D.2 that the new terms reflect the existence at the microscopic level of a profile of stress across the thickness of the membrane whose first moment is the torque density, Eq. (2.61). The Helfrich stress tensor Eq. (2.64) contains the expected component Eq. (2.63) normal to the membrane plane. The remaining in-plane stress can be resolved into components perpendicular to and along the boundary,

$$T_\perp = \mathbf{T}(\hat{\mathbf{p}}) \cdot \hat{\mathbf{n}} = p^\alpha \partial_\alpha \mathcal{M},$$

$$T_{\hat{\mathbf{p}}} = \mathbf{T}(\hat{\mathbf{p}}) \cdot \hat{\mathbf{p}} = \left(\tau_0 + \frac{\kappa_b}{2}(2H - C_0)^2\right) - \mathcal{M}p^\alpha C_\alpha^\beta p_\beta,$$

$$\mathbf{T}(\hat{\mathbf{p}}) \cdot (\hat{\mathbf{n}} \times \hat{\mathbf{p}}) = \mathcal{M}p^\alpha C_\alpha^\beta \epsilon_{\beta\gamma} p^\gamma = \sqrt{g}\mathcal{M}p^\alpha\left(C_\alpha^1 p^2 - C_\alpha^2 p^1\right), \tag{2.66}$$

where we have used notation and results from Appendix C. Equations (2.66) show clearly that the stresses are generally non-uniform, anisotropic and dependent on the local curvature. These expressions are particularly simple when $\hat{\mathbf{p}}$ and $(\hat{\mathbf{n}} \times \hat{\mathbf{p}})$ lie along the principal axes of the curvature tensor, in which case the second (shear) term vanishes and $p^\alpha C_\alpha^\beta p_\beta = C_{\hat{\mathbf{p}}}$, where $C_{\hat{\mathbf{p}}}$ is the principal curvature in the direction $\hat{\mathbf{p}}$. For other directions of $\hat{\mathbf{p}}$, the shear term is generally non-zero. Note from Eq. (2.66) that, when the principal curvatures are equal, $C_1 = C_2$, so $\mathbf{C}$ is proportional to the unit matrix, then the component of $\mathbf{T}(\hat{\mathbf{p}})$ along the boundary vanishes and the component $T_{\hat{\mathbf{p}}}$ along the in-plane normal becomes isotropic. These symmetry properties reflect the fluidity of the membrane.[28] In particular, the vanishing of $\mathbf{T}(\hat{\mathbf{p}}) \cdot (\hat{\mathbf{n}} \times \hat{\mathbf{p}})$ reflects the absence of shear rigidity in the fluid phase. For situations for which $C_1 \neq C_2$, the membrane does support a (small) shear stress proportional to the bending moment $\mathcal{M}$.

It is now straightforward to calculate the generalized divergence $D_\alpha \mathbf{T}^\alpha$ (see Appendix D, Eqs. (2.197) and (2.198)) and to evaluate the condition, Eq. (2.53), for force equilibrium. Working through the differential geometry gives, finally,

---

[28] Note, in contrast, that $T_\perp$ depends on in-plane derivatives of the curvature and does not vanish unless these are independent of direction.

$$\Delta P = 2H\left(\tau_0 + \frac{\kappa_b}{2}(2H - C_0)^2\right) - \Delta\mathcal{M} - \mathcal{M}(4H^2 - 2\det\mathbf{C})$$

$$\partial_\alpha \tau_0 = 0, \tag{2.67}$$

which we shall refer to as the Ou-Yang equations. These equations generalize Eqs. (2.46) and (2.47). Note that τ_0 remains uniform. The equation for the pressure difference was first derived by Ou-Yang and Helfrich (1987a) using a variational technique (see also Ou-Yang and Helfrich (1987b) and Ou-Yang and Helfrich (1989)). In the special case of a closed vesicle, where the area constraint is incorporated via the Lagrange multipliers ΔP and Σ, Eq. (2.38), it turns out that $\tau_0 = \Sigma$. Equilibrium membrane shapes of fluid membranes can in principle be obtained by solving Eq. (2.67) subject to appropriate boundary conditions and taking into account the self-consistency discussed at Eqs. (2.40) and (2.41). When there are no open boundaries, as is the case for vesicles, then suitable boundary conditions are provided by the requirements of fixed overall area and volume. When open boundaries are present, then the boundary stresses $\mathbf{T}(\hat{\mathbf{p}})$ and torques $\mathcal{M}$ required for equilibrium may be calculated by applying Eqs. (2.61) and (2.64) at the boundaries. These boundary stresses and torques are required for overall translational and rotational equilibrium, Eqs. (2.49) and (2.57), and must be supplied in the laboratory by some external agency, for example, the material of the aperture boundary.

In practice, integrating Eq. (2.67) has not proved to be a useful technique for finding equilibrium membrane shapes except for axisymmetric geometries. We can illustrate these ideas by applying the force analysis to the simple axisymmetric geometry shown in Fig. 2.12, which shows a uniform membrane section, closed at one end and open at the other. Axisymmetry guarantees that the forces in radial directions cancel, so that any net force must be axial. The net force to the left due to the pressure difference ΔP is $\pi r^2 \Delta P$. In order for equilibrium to hold, this force must be balanced by the boundary force along the length $2\pi r$ of the open lip. The force per unit length on the boundary has a component $\mathbf{T}_\perp$ in the normal direction $\hat{\mathbf{n}}$ and a component $\mathbf{T}_{\hat{\mathbf{p}}}$ in the plane of the membrane and perpendicular to the edge, as shown. It follows that

$$\pi r^2 \Delta P = 2\pi r\left(T_{\hat{\mathbf{p}}} \sin\theta - T_\perp \cos\theta\right), \tag{2.68}$$

which generalizes the argument based on Fig. 2.8 which lead to Eq. (2.45). The force densities are given in this case by Eq. (2.66), so $T_{\hat{\mathbf{p}}} = \left(\tau_0 + \frac{\kappa_b}{2}(2H - C_0)^2\right) - \mathcal{M}C_{\hat{\mathbf{p}}}$ and $T_\perp = \frac{d\mathcal{M}}{ds}$, where s is the arclength coordinate. It is convenient to use the relation $\frac{dr}{ds} = \cos\theta$ to cast the final result in the form,

$$\frac{d\mathcal{M}}{dr} = \frac{1}{\cos^2\theta}\left[\sin\theta\left(\tau_0 + \frac{\kappa_b}{2}(2H - C_0)^2 - \mathcal{M}C_{\hat{p}}\right) - \frac{r}{2}\Delta P\right]. \tag{2.69}$$

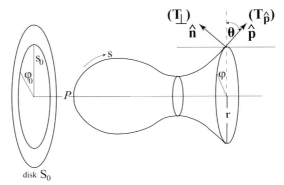

Fig. 2.12 Notation for describing an axisymmetric membrane patch. As shown, r measures radial distance from the rotation axis; s measures distance from the pole P along a line of longitude; θ measures the deviation of the tangent from the radial direction and φ is the azimuthal angle. $\hat{\mathbf{n}}$ is the unit outward normal vector; $\hat{\mathbf{p}}$ is the unit vector in the tangent plane perpendicular to the local element of membrane boundary. The stress-tensor components $\mathbf{T}_\perp$ and $\mathbf{T}_{\hat{\mathbf{p}}}$ act along $\hat{\mathbf{n}}$ and $\hat{\mathbf{p}}$, respectively. The elastic cytoskeletal disk S_0 shown at the left is discussed in Appendix D.6. The center of the relaxed disk is made coincident with the pole P and the remainder of the disk is deformed in an axisymmetric manner ($\varphi = \varphi_0$) to fit the membrane shape. The resulting elastic deformation of the disk is described in terms of the mapping $s(s_0)$.

This result, which we refer to as the Helfrich equation, was originally derived variationally by Deuling and Helfrich (1976) and is given in the historical notation in Appendix D.6 as Eq. (2.230). It expresses the axial force-balance condition and has been the basis of most axisymmetric vesicle-shape calculations. Note that this is a first-order differential equation and is clearly different from the Ou-Yang equation (2.67). The relation between them is that Eq. (2.69) is a first integral of Eq. (2.67) incorporating the condition that the net axial force is zero.[29]

Up to this point, we have ignored the Gaussian contribution to the energy, F_g, Eq. (2.15). Although we know that such a term cannot contribute to the shape determination of closed vesicles because of the Gauss–Bonnet theorem (see Section 2.3.2), it is generically present (Appendix B). Its effect on the equilibrium mechanics is discussed briefly in Appendix D.3. It turns out that it does not contribute to the stress tensor, so Eq. (2.64) remains unchanged; however, it does contribute an additional term to the linear density $\mathbf{N}(\hat{\mathbf{p}}) \equiv p_\alpha \mathbf{N}^\alpha \equiv p_\alpha N^{\alpha\beta} \mathbf{Y}_\beta$ of intrinsic boundary torque. Previously, with F_{sc} only, this torque density was $\mathbf{N}_{sc}(\hat{\mathbf{p}}) = \mathcal{M}(\hat{\mathbf{n}} \times \hat{\mathbf{p}})$, Eqs. (2.61) and (2.62), directed along the boundary and isotropic in magnitude. The new Gaussian contribution takes the form,

[29] If the axisymmetric membrane is under an overall non-zero axial tension, then that force must be added as a constant term to Eq. (2.68). This occurs, e.g., for cylindrical membranes under tension (Bukman et al. 1996).

$$\mathbf{N}_g^\alpha = \kappa_g g^{\alpha\gamma}\left(2H\delta_\gamma^\beta - C_\gamma^{\;\beta}\right)\hat{\mathbf{n}} \times \mathbf{Y}_\beta. \tag{2.70}$$

It is easy to show that this new contribution satisfies $D_\alpha \mathbf{N}_g^\alpha = 0$, so it integrates to zero around any closed boundary (via Gauss's law) and does not contribute to $\mathbf{N}_{\text{int}}$, Eq. (2.62), and does not affect Eq. (2.63). Interestingly, the local torques which are produced are not isotropic and have, in general, a component along $\hat{\mathbf{p}}$ in addition to one along $\hat{\mathbf{n}} \times \hat{\mathbf{p}}$,

$$\hat{\mathbf{p}} \cdot \mathbf{N}_g(\hat{\mathbf{p}}) = \kappa_g \left(2H\delta_\alpha^\beta - C_\alpha^{\;\beta}\right)\epsilon_{\beta\gamma}p^\gamma p^\alpha \tag{2.71}$$

$$(\hat{\mathbf{n}} \times \hat{\mathbf{p}}) \cdot \mathbf{N}_g(\hat{\mathbf{p}}) = \kappa_g \left(2H\delta_\alpha^\beta - C_\alpha^{\;\beta}\right)p_\beta p^\alpha. \tag{2.72}$$

These results are particularly simple in local Cartesian coordinates ($g_{\alpha\beta} = \delta_{\alpha\beta}$) aligned along the principal directions of the curvature tensor,

$$\hat{\mathbf{p}} \cdot \mathbf{N}_g(\hat{\mathbf{p}}) = \kappa_g \left(C_2 - C_1\right)p_1 p_2 \tag{2.73}$$

$$(\hat{\mathbf{n}} \times \hat{\mathbf{p}}) \cdot \mathbf{N}_g(\hat{\mathbf{p}}) = \kappa_g \left(C_1 p_2^2 + C_2 p_1^2\right). \tag{2.74}$$

Thus, if the edge happens to coincide locally with one of the principal axes, then the torque density is aligned along the edge and proportional to the principal curvature in the direction of the edge.

2.4.3
Mechanics of the Membrane Skeleton

The membrane skeleton, as we have modeled it in Section 2.3.3, consists of an infinitely thin isotropic elastic sheet without bending rigidity. The general stress-tensor analysis introduced in the previous section and leading to Eq. (2.53) applies to the membrane skeleton and now takes the form,

$$Q\hat{\mathbf{n}} + D_\alpha \mathbf{T}_{\text{ms}}^\alpha = \mathbf{0}, \tag{2.75}$$

where Q (which replaces the ΔP in Eq. (2.53)) is the the normal force per unit area exerted on the membrane skeleton by the plasma membrane (defined positive in the outward direction) and $\mathbf{T}_{\text{ms}}^\alpha$ is the surface stress tensor of the membrane skeleton. Now, $(T_{\text{ms}}^\alpha)_\perp \equiv 0$, since the membrane skeleton cannot (by hypothesis) support forces in the normal direction. Furthermore, $\mathbf{N}_{\text{int}} \equiv 0$, since the membrane skeleton cannot support intrinsic torques. Therefore, it follows from Eq. (2.60) that the remaining in-plane stress tensor $T_{\text{ms}}^{\alpha\beta}$ must be symmetric. Under these restrictions, it is easy to separate Eq. (2.75) into normal and in-plane components,[30]

$$Q = T_{\text{ms}}^{\alpha\beta} C_{\alpha\beta} = \operatorname{tr} \mathbf{C}\mathbf{T}_{\text{ms}}, \tag{2.76}$$

$$D_\alpha T_{\text{ms}}^{\alpha\beta} = \partial_\alpha T_{\text{ms}}^{\alpha\beta} + \Gamma_{\alpha\gamma}^\alpha T_{\text{ms}}^{\gamma\beta} + \Gamma_{\alpha\gamma}^\beta T_{\text{ms}}^{\alpha\gamma} = 0, \tag{2.77}$$

30) Note (Appendix C) that $D_\alpha \mathbf{T}_{\text{ms}}^\alpha = \left(D_\alpha T_{\text{ms}}^{\alpha\beta}\right)\mathbf{Y}_\beta - T_{\text{ms}}^{\alpha\beta} C_{\alpha\beta}\hat{\mathbf{n}}$.

where $\mathbf{C}$ is the curvature tensor and $\mathbf{T}_{\mathrm{ms}}$ is the stress tensor in the form $(T_{\mathrm{ms}})_\alpha^\beta$ (see Appendix C). Note that, unlike ΔP in Eq. (2.53), Q is not uniform but will in general vary from place to place on the membrane skeleton. In the special case for which the principal axes of $\mathbf{C}$ and $\mathbf{T}_{\mathrm{ms}}$ coincide, then

$$Q = \tau_1 C_1 + \tau_2 C_2 = \frac{\tau_1}{R_1} + \frac{\tau_2}{R_2}, \tag{2.78}$$

where $C_{1,2}$ and $\tau_{1,2}$ are the principal curvatures and the principal stresses, respectively. Equation (2.78) generalizes Eqs. (2.45) and (2.47). The two-component Eq. (2.77) expresses in-plane force equilibrium. The Christoffel symbols Γ, defined in Appendix C, encode the shape of the (plasma membrane) surface on which the elastic membrane (skeleton) is stretched.

The solution of Eqs. (2.76) and (2.77) determines the equilibrium stress-strain state of the membrane skeleton as it is stretched/compressed from its initial, undeformed shape $S_0 \equiv \mathbf{R}_0(s^1, s^2)$ to fit over the shape $\mathbf{R}(s^1, s^2)$ of the plasma membrane. In this context, Eq. (2.76) is a constraint equation which evaluates the necessary distribution of normal force Q acting on the elastic membrane to keep it localized at $\mathbf{R}$ (of course, there is a corresponding non-uniform pressure $-Q$ which acts on the plasma membrane). Once the form of the elastic constitutive relations is given, Eq. (2.77) determines the distribution of stress $T_{\mathrm{ms}}^{\alpha\beta}$ over the elastic membrane. Thus, for constitutive relations like Eqs. (2.27) or (2.28) which give the local elastic energy density $f_{\mathrm{ms}}(\lambda_1, \lambda_2)$ as a function of the local principal extension ratios, it follows immediately that the principal stresses are[31]

$$\tau_1 = \frac{1}{\lambda_2} \frac{\partial f_{\mathrm{ms}}}{\partial \lambda_1} \quad \text{and} \quad \tau_2 = \frac{1}{\lambda_1} \frac{\partial f_{\mathrm{ms}}}{\partial \lambda_2}, \tag{2.79}$$

so that in locally Cartesian coordinates oriented along the principal-axis directions $T^{\alpha\beta} = \begin{bmatrix} \tau_1 & 0 \\ 0 & \tau_2 \end{bmatrix}$. In simple cases the principal axes of $\mathbf{C}$ and $\mathbf{T}_{\mathrm{ms}}$ are aligned, and it is clear that Eq. (2.77) provides two equations to determine the distributions of the two unknowns λ_1 and λ_2. A situation of this type is worked out in Appendix D.6, where a circular patch S_0 is deformed to fit over an axisymmetric shape S. More generally, however, an additional angular variable $\theta(s^1, s^2)$ is required at each point to specify the unknown local orientation of the principal stress axes, so it might seem that there are only two equations to determine the three fields λ_1, λ_2 and θ. The resolution of this apparent paradox is that there is a symmetry relation (trivially satisfied in the axisymmetric case) due to the fact that the local deformation matrix $\mathbf{M} = \mathbf{M}(\lambda_1, \lambda_2, \theta)$ (see Appendix B) is determined as the in-plane part of $\partial \mathbf{R}_\alpha(s^1, s^2)/\partial (\mathbf{R}_0)_\beta(s^1, s^2)$. Thus, spatial derivatives of $\mathbf{M}$ are constrained

[31] The argument invokes the work theorem for an initial unit patch of membrane deformed to dimensions $\lambda_1 \times \lambda_2$ and then further to $(\lambda_1 + d\lambda_1) \times \lambda_2$. Thus, $\Delta E = \frac{\partial f_{\mathrm{ms}}}{\partial \lambda_1} d\lambda_1 =$ (force) $\times$ (distance) $= (\tau_1 \lambda_2) d\lambda_1$.

by the equality of cross derivatives in expressions like $\partial^2 \mathbf{R}_\alpha / \partial (\mathbf{R}_0)_\beta \partial (\mathbf{R}_0)_\gamma$, which provides another local relation between λ_1, λ_2 and θ. The general relation connecting the elastic energy density $f_{\mathrm{ms}}(\alpha, \beta)$, Eq. (2.26) and the elastic stress tensor of the membrane skeleton is

$$T_{\mathrm{ms}}^{\sigma\tau} = \left[\frac{\partial f_{\mathrm{ms}}}{\partial \alpha} - \left(\frac{1+\beta}{1+\alpha} \right) \frac{\partial f_{\mathrm{ms}}}{\partial \beta} \right] g^{\sigma\tau} + \frac{1}{(1+\alpha)^2} \frac{\partial f_{\mathrm{ms}}}{\partial \beta} g_0^{\sigma\tau}, \qquad (2.80)$$

where $g_0^{\sigma\tau}$ and $g^{\sigma\tau}$ are, respectively, the metric tensors of the undeformed and deformed surfaces, $\mathbf{R}_0(s^1, s^2)$ and $\mathbf{R}(s^1, s^2)$. This important result is proved in Appendix D.5. Equation (2.79) is a special case.

With the mechanics of the membrane skeleton now in hand, we are finally in a position to discuss the analytic approach to the coupled problem of the plasma membrane plus the membrane skeleton. The only effect of the cytoskeleton is to provide an additional (non-uniform) pressure $-Q$ on the plasma membrane, thus replacing $\Delta P \to \Delta P - Q$ in Eq. (2.53), which then passes through to the Ou-Yang equation (2.67). The in-plane forces are not affected, so τ_0 remains uniform. Note that the full red-cell calculation is now self-consistent: To calculate the shape of the plasma membrane from the Ou-Yang equation, one needs to know Q. To evaluate Q from Eq. (2.76) for the membrane skeleton, one needs to know the elastic stress tensor $T_{\mathrm{ms}}^{\alpha\beta}$ for the membrane skeleton (Eq. (2.80)) and the local curvature $\mathbf{C}$. To calculate the elastic stress tensor from Eq. (2.77), one needs to know the geometric quantities Γ. Finally, to find $\mathbf{C}$ and Γ, one needs to solve the plasma membrane shape problem. While this program has been carried out under the simplification of axisymmetry (Mukhopadhyay et al. 2002) (see also Appendix D.6), for more general shapes it has so far proven simpler to resort to numerical methods, as discussed in the next section.

2.5
Calculating Shapes Numerically

This section describes the technique we have used to find RBC shapes numerically for given values of the control parameters. We represent the membrane skeleton as a discrete triangular mesh of labeled vertices, first over the undeformed cytoskeletal shape S_0 and then, appropriately deformed, over the final membrane shape S. We will denote the undeformed and deformed meshes as $\widetilde{S}_0$ and $\widetilde{S}$, respectively. These meshes encode two kinds of information. On the one hand, they may be viewed as piecewise planar (triangulated) approximations to the shapes of the surfaces S_0 and S. On the other hand, the deformation of each triangular plaquette in the mapping from $\widetilde{S}_0$ to $\widetilde{S}$ encodes in discrete, approximate form the local elastic deformations of the membrane skeleton S_0 as it is distributed over the final membrane shape S. The calculation involves constructing the initial and final triangulations, representing the (free) energy $F[S_0; S]$, Eq. (2.6), in the discretized approximation $\widetilde{F}[\widetilde{S}_0; \widetilde{S}]$ (where we use the tilde to denote the discrete approximation), and then using Monte Carlo methods to minimize $\widetilde{F}$ by variation of $\widetilde{S}$. The final, minimizing mesh, $\widetilde{S}_{\min}$, then provides an approximate representation of the energy-minimizing RBC shape $S_{\min}$.

In very recent work Khairy et al. (2007) have numerically solved the same model we have used here by expanding in a spherical-harmonic representation. Details of this method have not yet been published; however, it is claimed to be effective and computationally economical, especially for rather smooth shapes.

2.5.1
Construction of an Initial Spherical Net $\widetilde{S}_{\text{sphere}}$

The initial discretized spherical surface, $\widetilde{S}_{\text{sphere}}$, is built up from an icosahedron by utilizing a standard method of constructing a sphere in computer graphics. Imagine an icosahedron inscribed within a sphere of radius R_A. This icosahedron, which constitutes the initial, level-zero triangulation, has 20 triangular faces, 12 vertices (all with five-fold coordination) and 30 edges. In each refinement of this triangulation, we replace each original equilateral triangle by four equal smaller equilateral triangles by projecting the midpoint of each edge onto the circumscribing sphere and then connecting those points to the three original vertices and to one another. In this process the number of edges, N_e, and the number of triangular faces ("plaquettes"), N_t, both increase by four times and so, after N_{div} generations, $N_e = 30 \times 4^{N_{\text{div}}}$ and $N_t = 20 \times 4^{N_{\text{div}}}$. The number of vertices N_v is then determined by Euler's theorem (Millman and Parker 1977, p. 188): $N_v = N_e - N_t + 2 = 10 \times 4^{N_{\text{div}}} + 2$. Note that the 12 initial vertices retain their five-fold connectivity, whereas all the others have a six-fold connectivity. The former are referred to as defective vertices and the latter, as regular vertices.

Information pertaining to the mesh is stored in three files. The first contains the vertex coordinates calculated to a precision of 16 decimal places; the second, the vertex indices of each triangle, ordered anti-clockwise as seen from the outside; the third, the indices of the five or six nearest-neighbor vertices of each vertex, also ordered anti-clockwise as seen from the outside. The information within these files completely specifies the group of nearest and second-nearest triangles about each vertex, which we refer to as a cluster. As illustrated in Fig. 2.13, there are three distinct types of clusters, depending on whether the central vertex or one of its nearest neighbors is defective. The geometries of these triangles are required for calculating the change in energy caused by a Monte-Carlo move of the central vertex k.

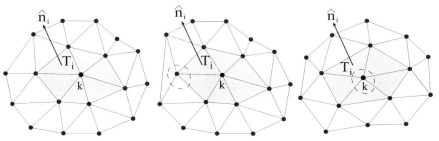

Fig. 2.13 Local plaquette clusters about a central vertex k, showing nearest- and next-nearest-neighbor triangles. When the central vertex undergoes a Monte Carlo move, the geometry of the full cluster is needed to compute the energy change $\Delta \widetilde{F}$. Each triangle T_i is associated with its outward normal $\hat{\mathbf{n}}_i$. Nearest-neighbor triangles are shown in grey. In the left-hand figure, all vertices are six-fold coordinated and the full cluster contains 24 triangles. In the central figure, one nearest-neighbor vertex (circled) is defective, so there are 23 triangles in the full cluster. In the right-hand figure, the central vertex k (circled) is defective and the full cluster contains 20 triangles.

This mesh refinement method is applied subsequently to $\widetilde{S}_0$ and $\widetilde{S}$ but without projecting the newly created vertices of $\widetilde{S}_0$ and $\widetilde{S}$ onto S_0 and S, respectively, since S_0 (during its construction) and S are not known a priori. In practice, the number of mesh refinements required for sufficient numerical accuracy (Lim 2003) is $N_{\text{div}} = 3$ ($N_t = 1280$) for surfaces with smooth morphologies, for example, axisymmetric and non-axisymmetric discocytes, stomatocytes I and knizocytes, and $N_{\text{div}} = 4$ ($N_t = 5120$) for surfaces with sharper features, for example, echinocytes I, II and III and stomatocytes II and III.

In the remaining subsections we will need a notation for describing the mesh and its deformation. Thus, each initial triangular plaquette T^0 of $\widetilde{S}_0$ transforms into a plaquette T of $\widetilde{S}$, as illustrated in Fig. 2.14. In this process the edge vectors $\mathbf{l}_0$ and $\mathbf{l}'_0$ of T^0 transform into edges $\mathbf{l}$ and $\mathbf{l}'$ of T. We will assume that the vectors $\mathbf{l}$ and $\mathbf{l}'$ are chosen so that $\mathbf{N} \equiv \mathbf{l} \times \mathbf{l}'$ points outwards. The unit outward normal vector to T is $\hat{\mathbf{n}} = \mathbf{N}/|\mathbf{N}|$. In the following, sums over plaquettes, edges and vertices will all appear. We adopt the convention of distinguishing such sums from one another by labeling them with indices i, j and k, respectively.

2.5.2
Discretization of $\mathbf{F}_{\text{con}}[\mathbf{S}]$

It is convenient to enforce the area and volume constraints (Sections 2.1.3 and 2.3.1) by means of Eq. (2.8) using, however, the computational moduli K_A^* and K_V^* of Table 2.2 in place of the physical moduli K_A and K_V, as discussed in Section 2.3.1.

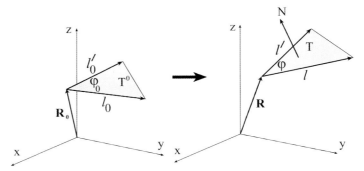

Fig. 2.14 Deformation of a representative triangular plaquette from its initial unstressed shape T^0 (as part of $\widetilde{S}_0$) to its final stressed shape T (as part of $\widetilde{S}$) with corresponding changes of the edge vectors $l_0 \to l$ and $l'_0 \to l'$ and of the included angle $\phi_0 \to \phi$. $\mathbf{N}$ is the cross product of l and l' and points in the direction of the unit normal $\hat{\mathbf{n}}$ to the plane of T.

The first step in approximating $F_{\text{con}}[S]$ is to approximate the area and volume, $A[S]$ and $V[S]$, by those of the equivalent mesh $\widetilde{S}$. Elementary algebra gives

$$\widetilde{A}[\widetilde{S}] = \sum_i \Delta A_i = \frac{1}{2}\sum_{i=1}^{N_t}|\mathbf{N}_i| \qquad (2.81)$$

and

$$\widetilde{V}[\widetilde{S}] = \frac{1}{6}\sum_{i=1}^{N_t}\mathbf{R}_i \cdot \mathbf{N}_i, \qquad (2.82)$$

where the sums run over the number N_t of triangular plaquettes and ΔA_i is the area of plaquette i. Once $\widetilde{A}$ and $\widetilde{V}$ are found using Eqs. (2.81) and (2.82), they are substituted into Eqs. (2.9) and (2.12),

$$\widetilde{F}_A[\widetilde{S}] = \frac{K_A^*\left(\widetilde{A}[\widetilde{S}] - A_0\right)^2}{2A_{\text{RBC}}} \qquad (2.83)$$

and

$$\widetilde{F}_V[\widetilde{S}] = \frac{K_V^*\left(\widetilde{V}[\widetilde{S}] - V_0\right)^2}{2V_{\text{RBC}}}, \qquad (2.84)$$

to give the approximate versions of $F_A[S]$ and $F_V[S]$, respectively, the sum of which provides the computational constraint,

$$\widetilde{F}_{\text{con}}[\widetilde{S}] = \widetilde{F}_V[\widetilde{S}] + \widetilde{F}_A[\widetilde{S}]. \qquad (2.85)$$

2.5.3
Discretization of $F_{pm}[S]$

The bending energy $F_{pm}[S]$ of the plasma membrane is given by Eq. (2.13) without the Gaussian term or, equivalently, by Eq. (2.22). These expressions contain two surface integrals over the mean curvature H of S for which we require discrete approximations,

$$I_H[S] = \oint_S dA\, H \longrightarrow \sum_{k=1}^{N_v} \Delta A_k \overline{H}_k = \widetilde{I}_H[\widetilde{S}] \qquad (2.86)$$

and

$$I_{H^2}[S] = \oint_S dA\, H^2 \longrightarrow \sum_{k=1}^{N_v} \Delta A_k \overline{H}_k^2 = \widetilde{I}_{H^2}[\widetilde{S}]. \qquad (2.87)$$

In the discrete expressions at the right the sum is over vertices k of the mesh, ΔA_k is the area element associated with each vertex and the expression $\overline{H}_k$ represents an appropriate local average of the mean curvature of S. The difficulty is that the discretized surface, $\widetilde{S}$, is piecewise planar and intrinsically lacks a well-defined curvature, so it is not clear how $\overline{H}_k$ is to be defined on $\widetilde{S}$. This is the problematic part of formulating a sensible discrete approximation to F_{pm}. We will discretize F_{pm} based on a method due to Jülicher (1996), which represents F_{pm} exactly in the continuum limit, unlike the commonly used earlier method of Kantor and Nelson (1987) (Kantor 1989), which is only approximate (Gompper and Kroll 1997).

What Jülicher (1996) (see also (Kern 1998)) showed is that, in a way that becomes exact in the continuum limit (finer and finer mesh), the curvature may be regarded as concentrated at the sharp edges between the triangular plaquettes. Thus, in Fig. 2.15, two adjacent triangular plaquettes meet along an edge *l* which has been rounded into a cylindrical surface of radius r and length l. The mean curvature, Eq. (2.17), is zero everywhere except on the cylindrical surface, where it has the constant value $H = 1/2r$. The area of the exposed cylindrical surface is $lr\theta$, where θ is the angle between the plaquette normals $\hat{\mathbf{n}}$ and $\hat{\mathbf{n}}'$, so $\hat{\mathbf{n}} \cdot \hat{\mathbf{n}}' = \cos\theta$. It follows that the integrated mean curvature over the entire configuration of the two adjacent plaquettes is the product $\frac{1}{2}l\theta$, in which the cylinder radius has cancelled out, so we can take the limit $r \to 0$, recreating the sharp edges of $\widetilde{S}$. This limit provides a clean definition of the discretized integral Eq. (2.86),

$$\widetilde{I}_H[\widetilde{S}] = \frac{1}{2}\sum_{j=1}^{N_e} l_j \theta_j = \frac{1}{4}\sum_{k=1}^{N_v}\left(\sum_{j_k} l_{j_k}\theta_{j_k}\right), \qquad (2.88)$$

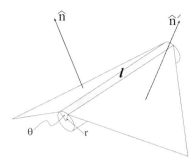

Fig. 2.15 Two plaquettes meet along the edge *l*, here shown expanded into a cylinder which the flat triangles meet tangentially along the two parallel lines shown. The turning angle θ which these two lines subtend along the cylinder axis is the same as the angle between the plaquette normals $\hat{n}$ and $\hat{n}'$. The integrated mean curvature of this plaquette geometry is treated as concentrated on the cylindrical surface and has a well-defined value in the limit $r \to 0$. It is shown by Jülicher (1996) that the integrated mean curvature of $\widetilde{S}$ calculated in this way agrees with that of the smooth surface S in the limit as the triangulation mesh becomes fine.

where the sum on j_k in the right hand term is over the edges[32] incident at the vertex k and the extra factor of $1/2$ is due to the fact that each edge is incident at two vertices. Comparing Eqs. (2.86) and (2.88) suggests that we define,

$$\overline{H}_k \equiv \frac{1}{4\Delta A_k} \sum_{j_k} l_{j_k} \theta_{j_k}. \tag{2.89}$$

It is convenient to construct the area element ΔA_k to be associated with each vertex by taking $1/3$ the area of each of the surrounding plaquettes, so

$$\Delta A_k \equiv \frac{1}{3} \sum_{i_k} \Delta A_i, \tag{2.90}$$

where the sum i_k is over the six triangles (five for defective vertices) which share the vertex k, as illustrated in Fig. 2.16.

Although plausible, the prescriptions, Eqs. (2.89) and (2.90), do not prove that $\overline{H}_k$ approaches the local mean curvature of S in the neighborhood of k as the triangulation becomes infinitely fine (provided, of course, that S is locally smooth). It is tedious but not difficult to persuade oneself that this is, indeed, the case by working close to the limit, where all plaquettes near k are very close to the tangent plane, so the Monge representation is convenient and expansion techniques can be used.

Using these results, we arrive finally at the discretized approximation to the bending energy F_{pm}, Eq. (2.13), of the plasma membrane (the Gaussian term has been dropped),

[32] Note that the number of terms in the sum is six for regular vertices and five for defective ones.

Fig. 2.16 The area ΔA_k associated with vertex k (shown shaded) is precisely one third of the total area of the triangular plaquettes surrounding k. The dotted lines shown connect each vertex to the midpoint of the opposite side. The three such lines for each triangle meet at the centroid.

$$\widetilde{F}_{\text{pm}}[\widetilde{S}] = 2\kappa_b \widetilde{I}_{H^2}[\widetilde{S}] - \frac{\overline{\kappa}\,\overline{m}_0}{R_A} \widetilde{I}_H[\widetilde{S}] + \frac{\overline{\kappa}}{2R_A^2} \widetilde{I}_H^2[\widetilde{S}], \tag{2.91}$$

where Eqs. (2.1), (2.23) and (2.33) have been used.

2.5.4
Discretization of $\mathbf{F_{ms}}$

In the mesh representation we are using, the deformation of the membrane skeleton in going from the relaxed configuration, S_0, to the final configuration, S, is represented discretely by the deformation of the mesh in going from $\widetilde{S}_0$ to $\widetilde{S}$. In this process, each plaquette i goes from an initial unstrained configuration, T_i^0, to a final strained configuration, T_i, as illustrated in Fig. 2.14. We take the the strain to be uniform over each plaquette, so discretization of the membrane skeletal stretch and shear energy Eq. (2.28) takes the form,

$$\widetilde{F}_{\text{ms}}[\widetilde{S}_0; \widetilde{S}] = \sum_{i=1}^{N_t} (\Delta A_0)_i \left(\frac{K_\alpha}{2} \left(\alpha_i^2 + a_3 \alpha_i^3 + a_4 \alpha_i^4 \right) \right.$$
$$\left. + \mu \left(\beta_i + b_1 \alpha_i \beta_i + b_2 \beta_i^2 \right) \right), \tag{2.92}$$

where the plaquette area $(\Delta A_0)_i$ refers to $\widetilde{S}_0$ and α_i and β_i are the area and shear strains, Eqs. (2.24) and (2.25), respectively, for plaquette i (assumed uniform).

It remains to calculate the strains associated with the deformation shown in Fig. 2.14. It is convenient to do this by finding the deformation matrix $\mathbf{M}$ (see Appendix B) which takes T^0 into T. There is no loss of generality in assuming that $\boldsymbol{l}_0$ and $\boldsymbol{l}$ are aligned, in which case it is easy to show that $\mathbf{M} = \begin{bmatrix} a & b \\ 0 & c \end{bmatrix}$, with

$$a = \frac{l}{l_0}, \tag{2.93}$$

$$b = \frac{1}{\sin\phi_0}\left(\frac{l'}{l'_0}\cos\phi - \frac{l}{l_0}\cos\phi_0\right), \tag{2.94}$$

$$c = \frac{l'}{l'_0}\frac{\sin\phi}{\sin\phi_0}. \tag{2.95}$$

It is then straightforward to calculate the invariants, Eq. (2.117),

$$\det \mathbf{M}^\mathbf{T}\mathbf{M} = a^2 c^2 \tag{2.96}$$

$$\operatorname{tr} \mathbf{M}^\mathbf{T}\mathbf{M} = a^2 + b^2 + c^2, \tag{2.97}$$

and to construct the strains, Eq. (2.24) and (2.25),

$$\alpha = ac - 1 = \frac{\Delta A}{\Delta A_0} - 1, \tag{2.98}$$

$$\beta = \frac{1}{2ac}\left(a^2 + b^2 + c^2 - 2ac\right). \tag{2.99}$$

This completes the calculation of the strains $\alpha(l_0, \phi_0, l, \phi)$ and $\beta(l_0, \phi_0, l, \phi)$ induced by the deformation of each triangular plaquette in going from $\widetilde{S}_0$ to $\widetilde{S}$. Substitution into Eq. (2.92) completes the evaluation of the discretized membrane-skeleton energy.

2.5.5
Energy Minimization by the Metropolis Monte Carlo Algorithm

At this point we have all the necessary preparation for calculating the discretized approximation of the total membrane free energy,

$$\widetilde{F}[\widetilde{S}_0; \widetilde{S}] = \widetilde{F}_{\text{con}}[\widetilde{S}] + \widetilde{F}_{\text{pm}}[\widetilde{S}] + \widetilde{F}_{\text{ms}}[\widetilde{S}_0; \widetilde{S}], \tag{2.100}$$

where the individual terms have been given in Eqs. (2.85), (2.91) and (2.92). In practice, the computational moduli K_A^* and K_V^* in $\widetilde{F}_{\text{con}}$, although softer than the physical moduli K_A and K_V (see Table 2.2), are stiff enough to fix the area and volume of $\widetilde{S}$ very close to $A_0 = A_{RBC}$ and $V_0 = V_{RBC}$, respectively. Under these conditions $\widetilde{F}_{\text{con}}$ is small and $\widetilde{F} \approx \widetilde{F}_{\text{m}}$, the discrete approximation to the membrane energy Eq. (2.7) of the red cell at fixed volume and area.

We proceed now to the final step in minimizing $\widetilde{F}$, implementation of the Metropolis algorithm to carry out the actual minimization (Allen and Tildesley 1989; Newman and Barkema 1999). For computational efficiency the twelve defective vertices and the $N_v - 12$ regular vertices are separated at the beginning by numbering the former as the first twelve followed by the latter in the array of all vertex indices. Each vertex on $\widetilde{S}$ is picked sequentially from the array and subjected to a trial move. In each such move, the vertex is displaced randomly by a small amount through the use of a pseudo-random number generator (the simple multiplicative congruential generator (Press et al. 1994)) to produce a slightly perturbed surface $\widetilde{S}'$ from $\widetilde{S}$. This vertex displacement gives rise to a change in the energy $\widetilde{F}$, defined by $\Delta\widetilde{F} \equiv \widetilde{F}[\widetilde{S}_0; \widetilde{S}'] - \widetilde{F}[\widetilde{S}_0; \widetilde{S}]$, which is calculated from the new and old geometries of the triangles of the cluster about the displaced vertex (see Fig. 2.13). Moving the central vertex of a cluster redefines only the nearest-neighbor (grey) plaquettes. Thus, in calculating the energy change $\Delta\widetilde{F}$ due to each move, only a small number of terms in $\widetilde{F}$ need to be considered. For $\widetilde{F}_{\text{con}}$ and $\widetilde{F}_{\text{ms}}$, Eqs. (2.85) and (2.92), changes occur only in the six terms i (five for the defective vertices) associated with the grey plaquettes. For $\widetilde{F}_{\text{pm}}$, Eq. (2.91), the sums Eqs. (2.86) and (2.87) defining $\widetilde{I}_H$ and $\widetilde{I}_H^2$ use $\overline{H}_k$, Eq. (2.89), which involves all the surrounding plaquettes and their edges j_k. It follows that moving the central vertex k affects contributions both from k and from its six (or five) nearest-neighbor vertices. Finally, the last part of each trial move is to accept the change if the Boltzmann factor $\exp(-\Delta\widetilde{F}/k_\mathrm{B}T)$ is greater than a randomly chosen number in the interval [0, 1] or, otherwise, to reject it and to revert back to the previous coordinates. By using a very low temperature T in the Boltzmann factor, trial moves that substantially raise $\widetilde{F}$ (positive $\Delta\widetilde{F}$) are heavily penalized, while moves that lower $\widetilde{F}$ (negative $\Delta\widetilde{F}$) are always accepted. The probability of acceptance of a trial move is kept at $(50 \pm 2)\%$ by adjusting the step size of the trial moves throughout each minimization run.

In the above, k_B is Boltzmann's constant and T is a user-defined minimization parameter, referred to as the "computational temperature." If T is chosen to be the physical temperature of the RBC physiological environment, then the Monte Carlo simulation will over time reproduce the full thermal ensemble of RBC shapes with its corresponding thermal distribution of energies F. Although this can be useful in interpreting experiments where thermal fluctuations are important (see Section 2.7.5), we shall for the most part be interested in constructing phase and stability diagrams and looking at the energy-minimizing shapes S_{min}. For these purposes, it is useful to set T to a value much lower than the physiological temperature, so that each local energy minimum can be followed independently.

A Monte Carlo run consists in starting with an initial configuration and running randomly through the vertices of the mesh. One trial per mesh vertex constitutes a "sweep." The total number of sweeps, N_{sweep}, is an important computational parameter. In practice, the energy $\widetilde{F}$ goes through an initial transient and then settles into a stationary random behavior with fluctuations which are small if T is low. We disregard data from the transient period and calculate the average $\langle \widetilde{F} \rangle_T$ over a sufficient number of sweeps to provide adequate statistics. Note that neither $\widetilde{F}_{\text{min}}$ nor $\widetilde{S}_{\text{min}}$ is directly attainable from the Metropolis algorithm, since the algorithm relies on the use of a non-zero T and, hence, cannot enforce strict minimization. However, a good approximation to both can be had by setting T to a reasonably low value, usually one tenth of room temperature or less (more in Section 2.6). The values of N_{sweep} and T used are dependent upon the size of the system (as measured by N_t, for example) and the type of operation under consideration. There are two types of operations: (i) the preparation of the set of reference surfaces $\widetilde{S}_0$ (Section 2.3.3) and (ii) the production of the minimum energy $\widetilde{S}$ for a given $\widetilde{S}_0$ (Section 2.6). The values of N_{sweep} and T for each will be given as the need arises.

In principle, one can vary T and N_t to obtain $\langle \widetilde{F} \rangle_T$ as a function of T and N_t and then extrapolate systematically to $T = 0$ and $N_t^{-1} = 0$ to find $S_{\min}$ and $F_{\min}[S_{\min}]$. We have done this in a few special cases. In practice, however, it is neither necessary nor feasible for the results presented here. At the low computational temperatures we employ, the extrapolations (when we can do them) result in only small changes. Furthermore, extrapolation on N_t is only feasible for smooth shapes. In practice, we can use N_t values of 1280 and 5120 but the next step up at 20 480 is prohibitively time consuming. Thus, we can only begin to extrapolate shapes which can be reasonably well approximated by the N_t=1280 mesh, which is OK for smooth shapes like the discocyte but excludes the highly spiculated echinocytes.

2.6
Predicted Shapes and Shape Transformations of the RBC

In this section we describe the results of numerically minimizing (see Section 2.5) the overall free energy functional $F[S_0; S]$, Eq. (2.6), of the RBC membrane for the parameter values listed in Table 2.2 and discussed in Appendix A. In general, we have treated as variable ("control parameters") only two parameters, the volume V_{ms} of the membrane-skeleton reference shape and the reduced effective spontaneous curvature (or reduced effective area difference), which is by convention described in terms of the variable $\overline{m}_0$ defined by Eq. (2.33). For each pair of values $(\overline{m}_0, V_{\text{ms}})$ there may be one or more local energy minima, corresponding to one or several mechanically stable membrane configurations. The lowest-energy minimizing shape is the putative zero-temperature ($T = 0$) ground state. Other minimizing shapes are metastable. The energies $F_{\min}$ of the minimizing shapes generally vary smoothly with the control parameters, thus forming a set of sheets $F_{\min}^{(\alpha)}(\overline{m}_0, V_{\text{ms}})$ indexed by the discrete superscript α. $S^{(\alpha)}(\overline{m}_0, V_{\text{ms}})$ denotes the minimizing shape associated with the sheet α at the phase point $(\overline{m}_0, V_{\text{ms}})$. The analogue of a thermodynamic phase diagram would be a plot over the $(\overline{m}_0, V_{\mathrm{ms}})$ plane of the ground-state shape class. Such a phase diagram can, indeed, be constructed; however, it does not generally correspond to what is seen in the lab. The reason is that the energy barriers between the minima are generically of order $\kappa_{\mathrm{b}} \sim 50\,\kappa_{\mathrm{b}} T_{\mathrm{room}}$, as discussed in Section 2.1.4, so metastable shapes are common in experiments.

As described generically in Section 2.1.6, distinct sheets can cross one another. In addition, one sheet can merge into or branch off from another along lines of mathematical bifurcation. Thus, each minimum-energy sheet α persists over some bounded region of the $(\overline{m}_0, V_{\mathrm{ms}})$ "phase space," which we call the region of stability of the "phase" or "shape class" α. The linear loci in $(\overline{m}_0, V_{\mathrm{ms}})$ which outline this region are called stability boundaries (or metastability boundaries) and constitute collectively the stability diagram of the phase α. The minimizing shape $S^{(\alpha)}(\overline{m}_0, V_{\mathrm{ms}})$ is the model's prediction for the RBC shape of class α at the phase point $(\overline{m}_0, V_{\mathrm{ms}})$. $S^{(\alpha)}(\overline{m}_0, V_{\mathrm{ms}})$ varies smoothly over the region of stability of the phase α; however, within a single phase, the symmetry and other generic features of the shape do not change. The shape classes α are closely related to the observational shape classifications (Bessis 1973) introduced in Sections 2.1.1 and 2.2.3, as we shall discuss in Section 2.6.1.

It is the aim of this section to describe the shapes, shape classes and stability diagrams which arise from energy minimization. Section 2.6.1 provides an overview of the minimizing shape classes and their relation to the Bessis (1973) classification illustrated in Figs. 2.2 and 2.3. Section 2.6.2 provides a generic discussion of the types of shape transitions that may be expected between different shape classes. Section 2.6.3 introduces the "phase-trajectory diagrams" which we constructed numerically in order to map out the various shape classes and their stability boundaries. Finally, in Section 2.6.4 we describe sequentially the 16 shape classes summarized in Table 2.3, their stability boundaries and their interrelations. With this information in hand, we go on in Section 2.7 to discussion of the significant results and predictions for RBC shapes and shape transitions which arise from this study.

2.6.1
Shape Classes

This section describes the nomenclature that we use to describe the shape classes that show up in our minimizations. Table 2.3 summarizes this nomenclature, introduces useful abbreviations, and provides links to illustrative figures. Some of these named classes refer to a single α sheet. Examples are the axisymmetric discocyte (AD) and the axisymmetric stomatocyte (AS). For reasons discussed below, others identify a collection or superclass of dis-

Table 2.3 Shape-class nomenclature. Each class is followed by its abbreviation and references to corresponding illustrative figures later in the text.

	Shape class	Abbreviation	Figure No.
1	Non-Axisymmetric Stomatocyte with triangular invagination	NAS-3	2.25
2	Non-Axisymmetric Stomatocyte with shallow invagination	NAS′	2.26
3	Non-Axisymmetric Stomatocyte	NAS	2.27
4	Axisymmetric Stomatocyte	AS	2.28
5	Axisymmetric Discocyte	AD	2.29 and 2.30
6	Non-Axisymmetric Discocyte	NAD	2.29 and 2.31
7	Echinocyte I with 9 bulges	E1-9	2.29, 2.32 to 2.34
8	Echinocyte I with 10 bulges	E1-10	2.35
9	Echinocyte I with 11 bulges	E1-11	2.36
10	Echinocyte I with 12 bulges	E1-12	2.37
11	Knizocyte	K	2.29, 2.38 and 2.39
12	Knizo-Echinocyte I, sub-class A	KE1-A	2.29 and 2.40
13	Knizo-Echinocyte I, sub-class B	KE1-B	2.41
14	Knizo-Echinocyte I, sub-class C	KE1-C	2.42
15	Spiculated Shape II	SS2	2.43
16	Spiculated Shape	SS	2.44

tinct but related sheets α. Examples are the non-axisymmetric stomatocytes (NAS, NAS′) and the spiculated shapes (SS and SS2).

Most, but not all, of these shapes occur as part of the normal SDE sequence, illustrated in Figs. 2.2 and 2.3; however, there is not a one-to-one correspondence between the traditional SDE terminology of Bessis (1973) and the largely symmetry-based nomenclature that we have adopted here. There are several reasons for this. First, the traditional terminology is partially qualitative, whereas the classes α that emerge from the minimization process are strictly defined and respect symmetry in a rigorous manner. Thus, classical stomatocytes I and II are both axisymmetric and differ only in the depth of the "cup." Both these shapes belong to the same axisymmetric stomatocyte (AS) sheet and in our terminology must, therefore, carry the same α label. By contrast, the stomatocyte III has a non-axisymmetric invagination; thus, because there is a symmetry change in passing from a stomatocyte II to any non-axisymmetric stomatocyte III (NAS) shape, these shapes belong to distinct minimization sheets α. The situation is further complicated because, as Fig. 2.3 illustrates, non-axisymmetric stomatocytes can have several distinct symmetries, thus corresponding to different sheets. Thus, there is not just one NAS sheet but, rather, a superclass of several distinct sheets of reduced symmetry. We lump these all these together as NAS and NAS′, each of which has several distinct α sub-classes. A different situation occurs for the echinocytic shapes. The weakly non-axisymmetric shapes called echinocyte I in the traditional terminology show up at $T = 0$ (see Section 2.7.5) with strictly periodic oscillations of the discocyte perimeter, that is, with an n-fold rotation axis at the position of the discocyte symmetry axis. Each of these periodicities corresponds to a distinct α sheet. We call them collectively echinocyte I (E1) but with an additional sublabel E1-n to distinguish the distinct periodicities. The echinocyte II and echinocyte III shapes exhibit complex patterns of spiculation, some symmetric but others with no symmetry. Each spicule pattern corresponds in the minimization to a separate sheet α. We have grouped these spiculated shapes (SS) into two broad groupings, SS and SS2, each of which contains many distinct sheets.

The classes summarized in Table 2.3 are distinguished principally by symmetry. The exceptions are NAS, NAS-3 and SS, which are grouped according to their respective surface features. NAS actually includes a few apparently symmetric shapes; however, it is hard to determine the values of $\overline{m}_0$ that separate these symmetric shapes from adjacent asymmetric ones and they all are discontinuously distinct from the main AS branch (see below). Therefore, we have chosen to group them together. NAS-3 and SS do not have obvious symmetries. In principle, shapes in the SS superclass could be divided further into sub-classes according to the number and arrangement of spicules; however, the number of such sub-classes is very large and the transitions between them are delicate and hysteretic, so we have chosen to group them together.

2.6.2
Shape Transitions, Trajectories and Hysteresis

Shape transitions, such as those of the SDE sequence, occur in the lab when some generalized applied force drives the red cell through a sequence of distinct shapes. From our point of view, the forcing occurs via changes in one or more of the control parameters which drive the mechanical system along a one-dimensional trajectory in its generalized phase space (e.g., along a line in $(\overline{m}_0, V_{\mathrm{ms}})$ in our example). If the driving is very fast, then the resulting time-evolution of RBC shapes can only be described by dynamics, and the $T = 0$ stable shape branches or sheets described here have no immediate relevance. If the driving is very slow, then what is observed at any $T > 0$ is a slowly varying shape ensemble, in which all shapes – stable and unstable – contribute according to their thermal weight. However, on appropriate intermediate time scales (which are relevant in the lab!) and at sufficiently low temperature, a sequence of identifiable shapes, each belonging to a specific sheet, may be seen. Such trajectories were previously described in Section 2.1.6, where we also introduced the distinction between continuous (ct) and discontinuous (dct) shape changes.

Our numerical shape calculations were carried out in a manner which closely parallels this experimental situation. Once we found, at a particular point $(\overline{m}_0, V_{\mathrm{ms}})$, a stable shape $S^{(\alpha)}(\overline{m}_0, V_{\mathrm{ms}})$, we would trace out the nearby parts of the sheet α by moving to a nearby point $(\overline{m}_0 + \delta\overline{m}_0, V_{\mathrm{ms}} + \delta V_{\mathrm{ms}})$ and using $S^{(\alpha)}(\overline{m}_0, V_{\mathrm{ms}})$ as the starting configuration in the energy-minimization routine. Note that our minimization technique is based on Monte Carlo simulation (see Section 2.5), so it has its own computational temperature.

In the remainder of this section we describe what we see computationally (and what we expect to observe in the lab) for trajectories which cross various kinds of phase and stability boundaries. Figure 2.17 summarizes what happens at several generic types of boundaries both at $T = 0$ and at $T > 0$. The horizontal axis represents distance along the driven trajectory. On the left side ("Discontinuous Transitions"), two distinct energy branches (sheets) cross, one or both of which terminate at an instability (i). Thus, at $T = 0$ in the top example at the left, if one starts out on A where it is the lower branch, the observed shape continues to change smoothly through the level crossing right up to the instability, at which point there is a sudden jump to a branch-B shape (well after the point where the energies crossed) along with a corresponding decrease of energy. At $T > 0$ there is a region near the instability (i), indicated by the shading, where the energy barrier between branches A and B becomes comparable to $k_B T$. A trajectory starting out on A now drops to B at some intrinsically unpredictable point within the shaded region. Sudden changes of shape like this are diagnostic of points of mechanical instability of a metastable branch ("metastability boundaries") and correspond to discontinuous (dct) phase boundaries.[33] If, on the other hand, the system starts on the higher branch (middle example at the left), then there is no change in shape class at the level crossing. The provisional stability of the higher-energy branch is a example of metastability. Metastability effects lead to a difference in the shape-transition point when the trajectory is reversed, as already noted in Fig. 2.2, so that the sequence of shapes is different in the two directions ("hysteresis"). This kind of behavior is illustrated by the example at the left of Fig. 2.17.

By contrast, the right side of Fig. 2.17 ("Continuous Transitions") illustrates what happens where one sheet A bifurcates at a continuous phase boundary to form a second sheet (typically of lower symmetry) of lower energy. The dashed continuation of A beyond the continuous transition (c) is

---

[33] Of course, if the shaded region extends through the level crossing, then there will be a region around the crossing where the state of the system can only be described as a (restricted) thermal mixture of A and B branches.

2.6 Predicted Shapes and Shape Transformations of the RBC

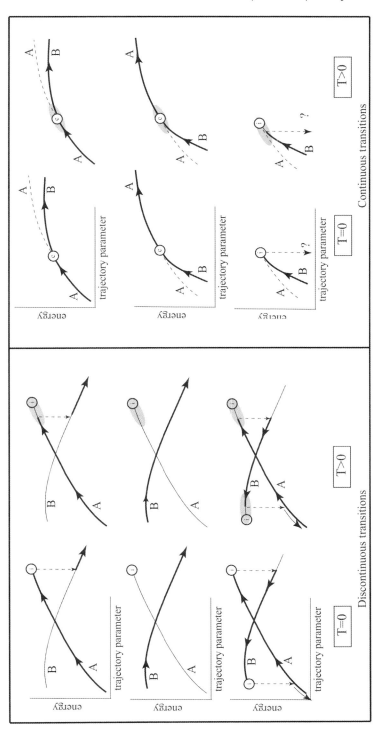

Fig. 2.17 Characteristic structures of level crossings (discontinuous transitions) and bifurcations (continuous transitions). Three examples of each type are shown. At discontinuous transitions (left), two distinct energy minima are separated by a thermal barrier, so that a shape in class A continues to evolve smoothly beyond the level crossing and does not change to class B until a later instability (i), resulting in hysteretic behavior. In contrast, at continuous transitions (right), shapes evolve smoothly through the bifurcation, which typically involves a breaking of symmetry. No hysteresis occurs; however, continuous transitions are characterized by a region (shaded) of enhanced thermal fluctuations at non-zero temperature.

typically an unstable branch, corresponding to a saddle of the energy surface. It is not accessible via Monte Carlo energy minimization. Thus, at $T = 0$ in the top example at the right, the shape passes smoothly from A to B at the transition. At $T > 0$ there is an interval (shaded) where an entire region of the energy surface around the A and B stationary points is thermally accessible. When the system passes through this region, one expects to see significant shape fluctuations, with representation in the thermal ensemble of shapes A and B as well as shapes intermediate between them. Regions of shape fluctuation are diagnostic of continuous (ct) phase boundaries. We will return to this situation later in Section 2.7.5 in discussing observed echinocyte I fluctuations. Although the change in symmetry at such a bifurcation occurs abruptly at $T = 0$, it is not so easy to pin down precisely in practice. Partly, this is because the change is continuous, so it starts out imperceptibly. More fundamentally, the fact that we are working at non-zero temperature both computationally and in the lab means that there is an intrinsic "fuzziness" corresponding to the shaded region. In the middle example at the right, the sense of the bifurcation has been reversed; again, the transition is smooth, with thermal fluctuations in the shaded region for $T > 0$. The bottom example at the right illustrates the annihilation of one stable and one unstable branch at an instability (i) followed by a sudden drop in energy to some lower branch. It is this kind of structure which is typically responsible for the instabilities following the level crossings at the left.

2.6.3
Phase-Trajectory Diagrams

We explored the sheets and stability boundaries computationally by fixing the volume V_{ms} of the skeletal reference shape S_0 and then stepping up and down incrementally in the effective-curvature (or effective-area-difference) variable $\overline{m}_0$, while monitoring the overall energy and the minimizing shape. In each step we used the previous minimizing shape as the starting point for the next minimization, thus following a single sheet as far as possible and mapping out a trajectory in the sense of the previous section. In this process we generally kept the computational temperature sufficiently low to make the fluctuation-dominated regions (shaded in Fig. 2.17) negligibly small. Stability boundaries were identified by locating the points in the trajectory at which the shape changed class. Discontinuous (dct) transitions were easily identified as abrupt shape changes accompanied by a jump to a lower overall energy. Continuous (ct) transitions were more difficult to pin down precisely, since they did not involve any sudden change of shape or energy.

Figures 2.18 to 2.24 summarize these results in what we refer to as "phase-trajectory diagrams" for a sequence of cytoskeletal volumes ranging from

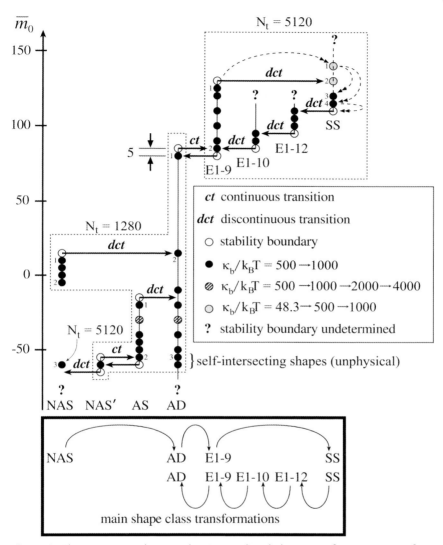

Fig. 2.18 Phase-trajectory diagram showing predicted shape transformations as a function of $\overline{m}_0$ at $V_{ms} = 100\ \mu m^3$. On the vertical axis, $\overline{m}_0$ is a dimensionless measure of the effective area difference between plasma-membrane leaflets. Shapes belonging to a given class are grouped sequentially along a single vertical line and their range in $\overline{m}_0$ indicates the region of stability of that class. There is no quantitative significance to the horizontal spacings; however, the horizontal arrows indicate the occurrence (and type) of shape-class transitions which take place as $\overline{m}_0$ is raised or lowered beyond the stability boundaries. Numerical labels on selected points key to specific calculated shapes shown in later figures. The inset below summarizes the sequence of predicted transitions as $\overline{m}_0$ increases (towards the right) or decreases (towards the left). Note the failure of reversibility through the E1-n regions, which is associated with hysteresis effects.

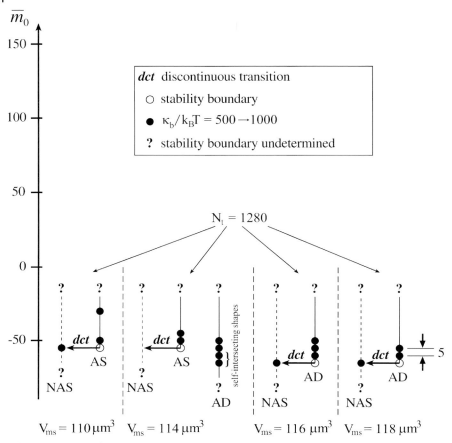

Fig. 2.19 Incomplete phase-trajectory diagrams for $\overline{m}_0$ at $V_{\rm ms} = 110$, 114, 116 and 118 μm^3. We took partial data at these values of $V_{\rm ms}$ only for the purpose of locating the lower stability boundaries of the NAS and AD shape-class phases. The standard step-size of $\Delta \overline{m}_0 = \pm 5$ is indicated at the right.

$V_{\rm ms} = 100$ μm^3 to $V_{\rm ms} = 155.8$ μm^3. Consider as a representative example Fig. 2.18. The vertical axis shows values of $\overline{m}_0$; the horizontal axis has no quantitative significance. Each vertical line indicates the range of $\overline{m}_0$ over which the corresponding phase is stable; thus, the endpoints of each line lie on stability boundaries. The vertical lines are solid for the axisymmetric phases and dashed for non-axisymmetric phases. They are labeled below according to the conventions of Table 2.3. In exploring these trajectories we used a step size in $\overline{m}_0$ of ± 5, which determines the uncertainty in the computed stability boundaries. Each circle indicates a point at which shape computations were carried out. Certain special points have been numbered for later reference. For example, the shapes corresponding to the AD points

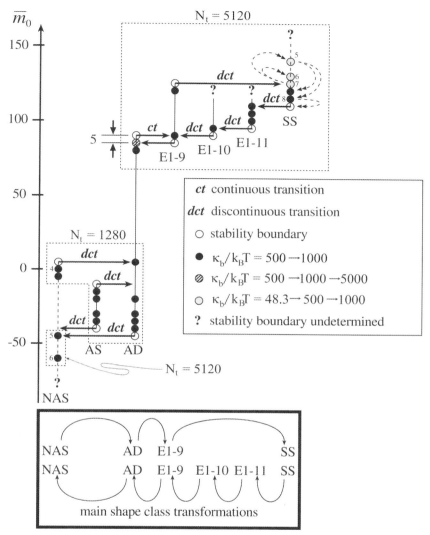

Fig. 2.20 Phase-trajectory diagram showing predicted shape transformations as a function of $\overline{m}_0$ at $V_{ms} = 130\ \mu m^3$. See text and caption of Fig. 2.18 for further information.

marked 1, 2 and 3 are shown in Fig. 2.30. The shading of the circles codes the range of computational temperatures used according to the key provided. The values of N_t indicate the fineness of the computational net (see below and Section 2.5). Stepping up or down along a single line, that is, on a single sheet, is reversible (no hysteresis). At the boundary of the range of stability of each shape class, a transition to another class occurs, as marked by the horizontal arrows which point to the new shape class and are labeled with

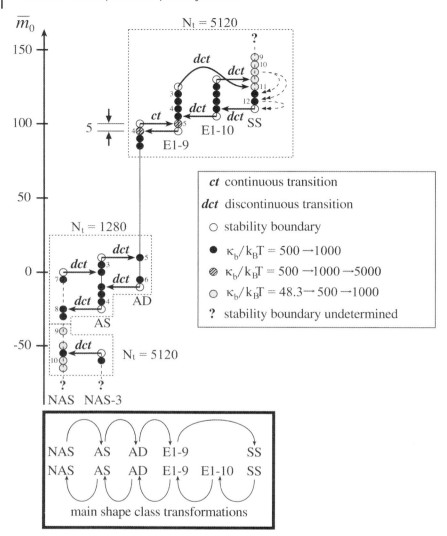

Fig. 2.21 Phase-trajectory diagram showing predicted shape transformations as a function of $\overline{m}_0$ at $V_{ms} = 148\ \mu m^3$. This value of $\overline{m}_0$ is our best estimate for the physical value; it was used in calculating the predicted erythrocyte shapes shown in Fig. 2.3. See text and caption of Fig. 2.18 for further information. Additional details of the AD to E1-n region are provided in Section 2.7.5 and Fig. 2.48.

the transition type (dct or ct). Thus, the range of the stable E1-9 phase terminates above in a discontinuous transition to a spiculated SS shape and below in a continuous transition to the axisymmetric discocyte (AD). On the other hand, the axisymmetric discocyte phase terminates below when the shapes become self-intersecting (unphysical) and above in a continuous transition

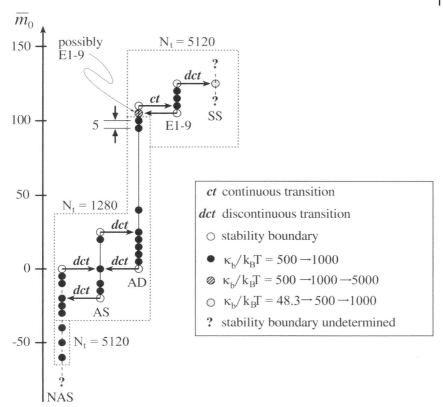

Fig. 2.22 Phase-trajectory diagram showing predicted shape transformations as a function of $\overline{m}_0$ at $V_{ms} = 152\,\mu m^3$. As indicated, one shape in the AD class near the upper stability boundary exhibited very slight undulations on the rim. However we have not grouped it under E1-9 because we were unable to exclude thermal fluctuations as the cause of the undulations (note that the corresponding boundary is continuous). See text and caption of Fig. 2.18 for further information.

to E1-9. The slight offsets of the two directions of the continuous transition is presumably a computational artifact associated with the finite step size, the difference in the mesh size, etc. Note that most transitions are discontinuous.[34] Note that the transition out of the spiculated SS class with decreasing $\overline{m}_0$ does not lead directly back to the E1-9 phase but, instead, passes through E1-12 and E1-10 in a cascade of discontinuous transformations. The box at the bottom of Fig. 2.18 summarizes the main shape-class transitions, specifically indicating hysteretic effects.

34) Standard Landau theory (Tolédano and Tolédano 1987; Chaikin and Lubensky 1995) shows that only transitions involving a symmetry which can be continuously broken are allowed to be continuous.

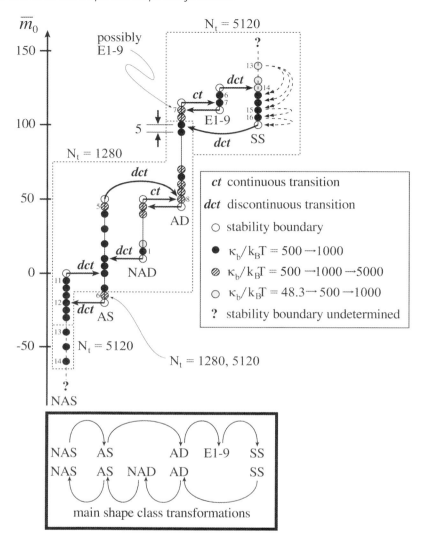

Fig. 2.23 Phase-trajectory diagram showing predicted shape transformations as a function of $\overline{m}_0$ at $V_{ms} = 154\,\mu m^3$. As indicated, one shape in the AD class near the upper stability boundary exhibited very slight undulations on the rim. However, we have not grouped it under E1-9 because we were unable to exclude thermal fluctuations as the cause of the undulations (note that the corresponding boundary is continuous). See text and caption of Fig. 2.18 for further information.

We have studied in some detail the cases $V_{ms} = 100, 130, 148, 152, 154$ and $155.8\,\mu m^3$, the results of which are reported in Figs. 2.18, 2.20, 2.21, 2.23 and 2.24, respectively. These figures illustrate the phase-stability behavior of all the major shape classes. In contrast, the cases $V_{ms} = 110, 114,$

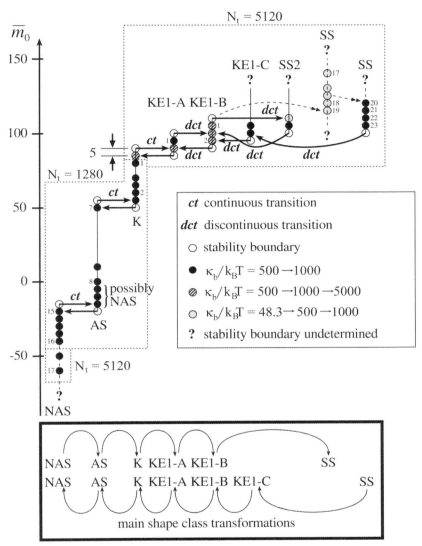

Fig. 2.24 Phase-trajectory diagram showing predicted shape transformations as a function of $\overline{m}_0$ at $V_{ms} = 155.8$ μm^3. As indicated, three shapes in the AS class near the lower stability boundary may be slightly non-axisymmetric. However, we have not grouped them under NAS because we were unable to exclude surface-triangulation roughness as the cause of this asymmetry. See text and caption of Fig. 2.18 for further information.

116 and $118\,\mu\mathrm{m}^3$, shown in Fig. 2.19, have only been explored sufficiently to allow determination of the lower stability boundaries of the AD and AS shape classes. Similarly, the case $V_{\mathrm{ms}} = 152\,\mu\mathrm{m}^3$, shown in Fig. 2.22, has only been explored sufficiently to allow determination of the boundaries of shape classes that are not spiculated.

Each point $(\overline{m}_0, V_{\mathrm{ms}})$ in Figs. 2.18–2.24 is associated with three computational parameters, N_{t}, $\kappa_{\mathrm{b}}/k_{\mathrm{B}}T$ and N_{sweep} (not shown), which are, respectively, the number of triangles used to approximate S, the normalized inverse computational temperature at which S is equilibrated, and the number of Monte Carlo sweeps. We have used either 1280 or 5120 triangles to represent S, except for one point in Fig. 2.23 where we have used both. Generally, we have chosen to use 1280 triangles to represent smooth S's and to use 5120 triangles to represent S's with sharper surface features. The highest equilibration temperature for each point is either $\kappa_{\mathrm{b}}/k_{\mathrm{B}}T = 500$ or $\kappa_{\mathrm{b}}/k_{\mathrm{B}}T = 48.3$ (room temperature). The former is used mainly in the determination of the regions of stability for the major non-spiculated shape classes. In the case of the SS and SS2 classes, it is used only in the determination of the lower stability boundaries. Exploration of the region where $\overline{m}_0 \geq 100$ is complicated by the occurrence of numerous locally stable spiculated shapes for the same V_{ms} and $\overline{m}_0$. In order to make our computation time feasible, we did not explore this region systematically. Instead, we limited our search to shapes at or near the absolute minima by equilibrating at $\kappa_{\mathrm{b}}/k_{\mathrm{B}}T = 48.3$ (room temperature) before raising $\kappa_{\mathrm{b}}/k_{\mathrm{B}}T$ to 500 and then to 1000. The shapes obtained in this manner belong to the spiculated shape classes (SS and SS2). In general, $N_{\mathrm{sweep}} \geq 2 \times 10^6$ for the equilibration at each temperature. Some parts of the NAS-3, NAS′, E1-10, E1-11, E-12 and KE1-C shape classes are not important for normal red-cell shapes and, thus, were not thoroughly explored.

Some data points in the SS class are associated with double-headed arrows. The tails of these arrows indicate the starting shapes. Those data points in the SS class not associated with double-headed arrows are the initial data that made us realize the complication arising from the multiplicity of locally stable spiculated shapes; we did not record the starting shapes of these initial data.

2.6.4
Individual Shape Classes and Stability Diagrams

In the following subsections we present the stability diagram and some characteristic shapes for each of the 16 shape classes. The stability boundaries may be read off from the phase-trajectory diagrams in the previous section. Full stability diagrams are given for all the major phases; for several minor phases we have only incomplete data. Figure 2.45 displays a composite of all the stability diagrams, so that their interrelations can be seen more clearly. The circles represent calculated shapes; the labeling is consistent with the phase-trajectory diagrams of Section 2.6.3. In some cases representative shapes are shown in separate figures; in others, they are shown on the stability diagram. Labeled shapes just outside the shaded stability range indicate the shape class which follows the instability. The order of presentation, from the most stomatocytic, through the discocytic and ending with the echinocytic phases, parallels the order in Table 2.3.

NAS-3

A locally stable non-axisymmetric stomatocyte with a triangular invagination is found at $V_{ms} = 148\,\mu m^3$ and $\overline{m}_0 = -60$, as shown in Fig. 2.25. It has no symmetry (the apparent three-fold axis is only approximate). The NAS-3 shape transforms discontinuously into the NAS class at the upper stability boundary. The NAS-3 shape is similar in appearance to the experimentally observed triangular stomatocyte shown in Fig. 2.3(B).

NAS'

A locally stable, shallow, non-axisymmetric stomatocyte with an ellipsoidal invagination is found at $V_{ms} = 100\,\mu m^3$ and $\overline{m}_0 = -60$, as shown in Fig. 2.26. It is mirror-symmetric with two mirror planes. This shape transforms continuously into the axisymmetric AS class at the upper stability boundary and discontinuously into the NAS class at the lower stability boundary.

NAS: Non-Axisymmetric Stomatocytes

The stability diagram and representative shapes NAS(1) to NAS(17) of the NAS class are shown in Fig. 2.27. NAS is a superclass consisting of a large collection of locally stable non-axisymmetric stomatocytic sheets with similar changes in surface features as $\overline{m}_0$ is varied, notably the narrowing and elongation of the initially oval invagination as $\overline{m}_0$ decreases. The NAS class includes some mirror-symmetric shapes with one or two mirror planes. We have not attempted to isolate these symmetric shapes from the non-symmetric ones because the continuous transformations between symmetric and non-symmetric shapes make it difficult to determine stability boundaries with precision. Shapes in the upper portion of the stability diagram, such as NAS(1), NAS(2), NAS(4), NAS(7), NAS(8), NAS(11), NAS(12) and NAS(15), resemble the stomatocyte II (c.f., Fig. 2.3(b)). Consider the shapes at $\overline{m}_0 = -60$, namely, NAS(3), NAS(6), NAS(10), NAS(14) and NAS(17). A comparison of these shapes reveals that only those with volumes in the range $130\,\mu m^3 < V_{ms} < 155.8\,\mu m^3$ exhibit the characteristic curved invagination

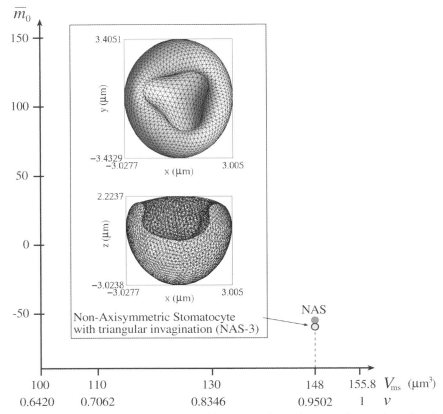

Fig. 2.25 Incomplete stability diagram of the NAS-3 shape class. v indicates the reduced volume (Eq. (2.3)) of the relaxed cytoskeletal shape S_0. Inset shows the locally stable shape found at $V_{ms} = 148\,\mu m^3$ and $\overline{m_0} = -60$. The three-fold rotation axis is only approximate. This shape bears a striking resemblance to the triangular stomatocyte shown in Fig. 2.3B and reported by Bessis (1972) (see also Section 2.2.3). At the next higher value of $\overline{m_0}$ the shape became NAS.

of the classic stomatocyte III (c.f., Fig. 2.3(a)). In other words, the reference shape S_0 must be highly inflated but not fully spherical in order for shapes resembling the stomatocyte III to occur. At the upper stability boundary, the NAS shape class transforms discontinuously into the axi- and up-down symmetric AD class, if $100\,\mu m^3 \leq V_{ms} < V_{AD/AS}$, or into the axisymmetric AS class, if $V_{AD/AS} < V_{ms} < 155.8\,\mu m^3$. The transition point $V_{AD/AS}$ is somewhere in the interval $130\,\mu m^3 < V_{ms} < 148\,\mu m^3$. The transformation into the AS class is continuous at $V_{ms} = 155.8\,\mu m^3$, which corresponds to a spherical S_0.

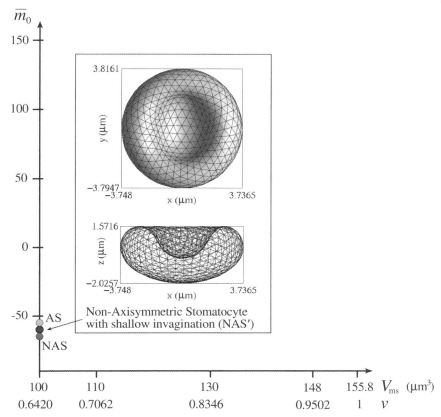

Fig. 2.26 Incomplete stability diagram of the NAS' shape class. Inset shows the locally stable shape found at $V_{ms} = 100\ \mu m^3$ and $\overline{m}_0 = -60$. The symmetry group contains two perpendicular mirror planes; the apparent axisymmetry is only approximate. At the next higher/lower values of $\overline{m}_0$ the shape became AS/NAS, respectively.

AS: Axisymmetric Stomatocytes

The stability diagram and representative shapes AS(1) to AS(8) of the AS shape class are shown in Fig. 2.28. The AS class consists of a single sheet of shapes which strongly resemble the classical stomatocyte I (c.f., Fig. 2.3(c)) and are characterized by possessing axisymmetry but lacking up-down symmetry. At the upper stability boundary, the AS class transforms discontinuously into the axi- and up-down symmetric AD class, except when S_0 is spherical, in which case the AS class transforms continuously into the K class. A notable feature in the transformation from AS to K is that the convex side of AS turns into the pinch of K, which corresponds to a change in shape from the axisymmetry of AS to the mirror symmetry of K. At the lower stabil-

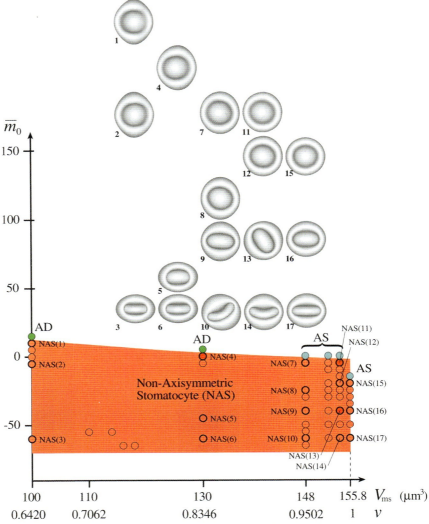

Fig. 2.27 Distribution of order parameter values as a function of temperature, at three cholesterol concentrations. Bi-modality in a plot reveals regions within the bilayer of different degrees of lipid order at that temperature and cholesterol concentration. (a) 5% cholesterol, (b) 25% cholesterol, (c) 3 % cholesterol. Bi-modality is seen at lower cholesterol concentration over a range of temperatures that decreases as the cholesterol concentration increases.

ity boundary, the AS class transforms discontinuously into the NAS class if $100\,\mu\text{m}^3 < V_{\text{ms}} < 155.8\,\mu\text{m}^3$, continuously into the mirror-symmetric NAS' class in the vicinity of $V_{\text{ms}} = 100\,\mu\text{m}^3$, and continuously into the NAS class if S_0 is spherical.

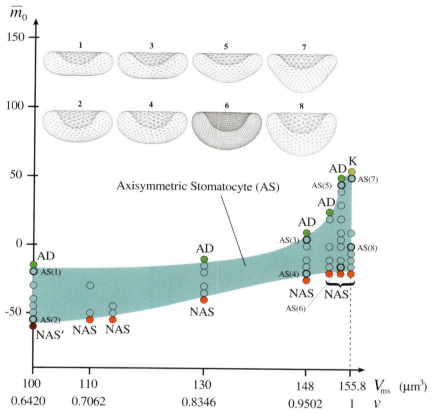

Fig. 2.28 Stability diagram of the AS shape class. The region of stability is shaded. The shape classes of selected points just outside the stability boundaries are indicated. Representative shapes AS(1) to AS(8) are shown above the stability region. The numbering is keyed both to the labeled points inside the stable region and to the corresponding points in phase-trajectory diagrams Figs. 2.18 to 2.24.

AD: Axisymmetric Discocytes

The stability diagram of the AD shape class is shown in Fig. 2.29. The representative shapes AD(1) to AD(8) are shown in Fig. 2.30. The AD class is characterized by axisymmetry plus up-down symmetry. It includes a subset of self-intersecting shapes located in the lower left corner of the stability diagram, where $V_{ms} \lesssim 114\,\mu m^3$. These self-intersecting shapes are not physical; they come about because we did not implement global self-avoidance of S. Upon crossing the lower stability boundary, AD transforms discontinuously into NAS, if $114\,\mu m^3 < V_{ms} < V_{NAS/AS}$, or into AS, if $V_{NAS/AS} < V_{ms} < 154\,\mu m^3$. The transition point $V_{NAS/AS}$ between these two regions is somewhere in the interval $130\,\mu m^3 < V_{ms} < 148\,\mu m^3$. AD trans-

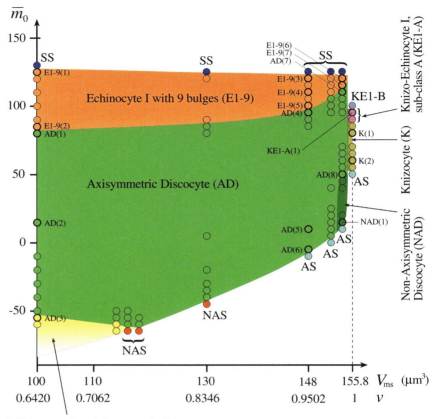

Fig. 2.29 Combined stability diagram of the shape classes AD, NAD, E1-9, K and KE1-A. The representative shapes AD(1) to AD(8), NAD(1), E1-9(1) to E1-9(7), K(1) and K(2) and KE1-A(1) are shown in Figs. 2.30, 2.31, 2.32 to 2.34, 2.38 and 2.39 and 2.40, respectively. The shape classes of selected points just outside the stability boundaries are indicated. Note that the transformation between AD and E1-9 and between AD and NAD are continuous.

forms continuously into the mirror- and up-down symmetric NAD with decreasing $\overline{m}_0$ in a narrow region about $V_{ms} = 154\,\mu\mathrm{m}^3$. AD transforms continuously into the 9-fold and up-down symmetric E1-9 at the upper stability boundary. As we did not investigate shape transformations in the narrow band defined by $154\,\mu\mathrm{m}^3 < V_{ms} < 155.8\,\mu\mathrm{m}^3$, we have no information on the shapes and the type of shape transformations that occur in that band; therefore, the continuous transformations depicted in Fig. 2.29 between AD and K and between AD and KE1-A are to be regarded as conjectural.

Figure 2.30 shows very clearly the progressive disappearance of the two dimples and the general flattening of the AD shape with increasing $\overline{m}_0$.

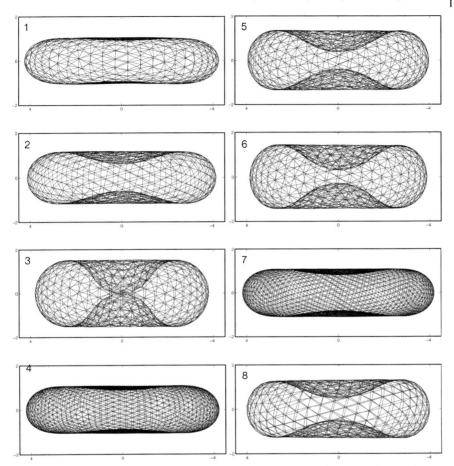

Fig. 2.30 Representative shapes AD(1) to AD(8) of the shape class AD. The numbering is keyed both to the stability diagram Fig. 2.29 and to the corresponding points in phase-trajectory diagrams Figs. 2.18 to 2.24.

Flattening of a discocytic RBC has also been observed experimentally in the shape transformation from a discocyte to an echinocyte I induced by a change in conformation of the transmembrane protein band 3 (Blank et al. 1994; Hoefner et al. 1994).

NAD: Non-Axisymmetric Discocytes

The stability diagram of the NAD shape class is shown in Fig. 2.29. The representative shape NAD(1) is shown in Fig. 2.31. The NAD class is characterized by up-down symmetry plus a single vertical mirror plane. The NAD stability diagram consists of a small region about $V_{\mathrm{ms}} = 154\,\mathrm{\mu m}^3$ that abuts

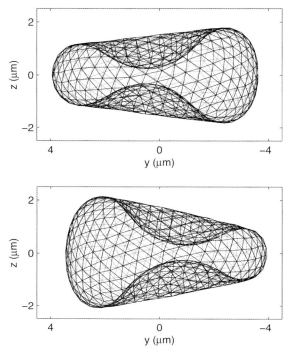

Fig. 2.31 Representative shape NAD(1) (upper) plus an NAD shape (lower) with doubled values of K_α and μ. Both shapes are at $V_{ms} = 154\ \mu m^3$ and $\overline{m}_0 = 15$. A laboratory image of an NAD shape is shown as Fig. 2.3A (see also Section 2.2.3).

the lower right corner of the much larger AD stability region at a line of continuous transitions. NAD transforms continuously into the axi- and up-down symmetric AD with increasing $\overline{m}_0$ and discontinuously into the axisymmetric AS at the lower stability boundary.

NAD(1) is similar in appearance to the experimentally observed non-axisymmetric discocyte shown in Fig. 2.3(A). Note, however, that the experimental shape does not occur naturally. The experimental shape is obtained by treating an osmotically swollen RBC with diamide and then osmotically shrinking it back to the normal volume (Fischer et al. 1981). The diamide treatment cross-links the spectrin tetramers of the membrane skeleton, which gives rise to two effects (Fischer et al. 1981): fixing S_0 close to a sphere and increasing the elastic constants K_α and μ. The fact that we find an NAD shape at lower (physiological) values of K_α and μ only when S_0 is nearly spherical suggests that a requirement for the production of a locally stable NAD is a nearly spherical S_0. In a separate, exploratory series of calculations using values of K_α and μ double those given in Table 2.2, we obtained another locally stable NAD shape quite

2.6 Predicted Shapes and Shape Transformations of the RBC

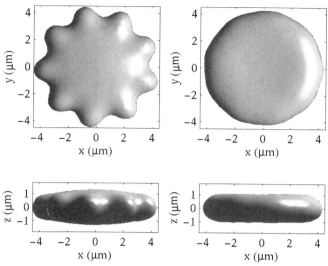

Fig. 2.32 Representative shapes E1-9(1) (left) and E1-9(2) (right) at $\overline{m}_0 = 125$ and 85, respectively and $V_{ms} = 100$ μm^3. Numbering keys to points in the stability diagram Fig. 2.29 and the phase trajectory diagram Fig. 2.18.

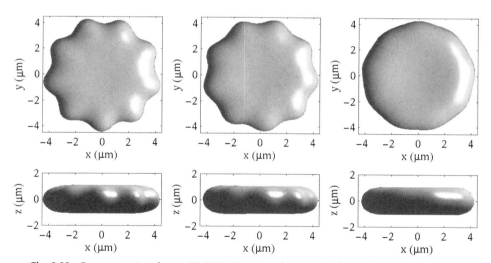

Fig. 2.33 Representative shapes E1-9(3), E1-9(4) and E1-9(5) (left to right) at $\overline{m}_0 = 120$, 110 and 100, respectively, and $V_{ms} = 148$ μm^3. Numbering keys to points in the stability diagram Fig. 2.29 and the phase trajectory diagram Fig. 2.21.

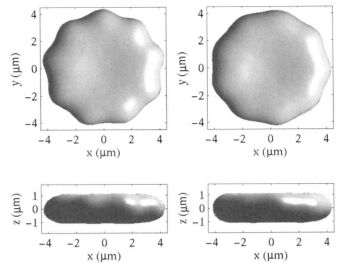

Fig. 2.34 Representative shapes E1-9(6) (left) and E1-9(7) (right) at $\overline{m}_0 = 120$ and 115, respectively, and $V_{\mathrm{ms}} = 154\,\mu\mathrm{m}^3$. Numbering keys to points in the stability diagram Fig. 2.29 and the phase trajectory diagram Fig. 2.23.

similar to NAD(1). This shape, shown together with NAD(1) in Fig. 2.31, has a more pronounced asymmetry in its rim thickness than NAD(1), suggesting that an increase in K_α and μ enhances the NAD shape asymmetry.

E1-9: Echinocytes I

The stability diagram of the E1-9 shape class is shown in Fig. 2.29. The representative shapes E1-9(1) to E1-9(7) are shown in Figs. 2.32 to 2.34. The E1-9 class is characterized by a 9-fold rotation axis plus up-down symmetry, reflecting the nine identical bulges that develop on the rim. The E1-9 shapes are effectively discocytes for which a 9-fold periodic undulation of the rim has broken the AD axisymmetry. The E1-9 stability diagram abuts the top of the AD stability diagram. E1-9 transforms continuously into the axi- and up-down symmetric AD with decreasing $\overline{m}_0$ and discontinuously into the SS class at the upper stability boundary. The range of $\overline{m}_0$ values over which E1-9 is locally stable decreases with increasing V_{ms} and eventually vanishes when S_0 becomes spherical.

E1-10, E1-11 and E1-12: Echinocytes I

Incomplete stability diagrams of the E1-10, E1-11 and E1-12 classes, each accompanied by a representative shape, are shown in Figs. 2.35 to 2.37, respectively. E1-10, E1-11 and E1-12 are characterized by 10-, 11- and 12-fold

2.6 Predicted Shapes and Shape Transformations of the RBC

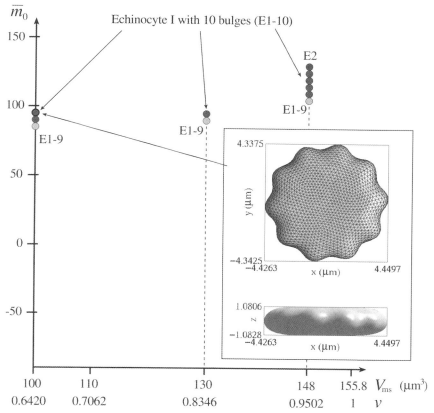

Fig. 2.35 Incomplete stability diagram of the shape class E1-10. Inset shows a representative shape at $V_{ms} = 100\ \mu m^3$ and $\overline{m}_0 = 95$. The shape classes of selected points just above or below the narrow range of E1-10 stability are indicated.

rotation axes, respectively, in addition to up-down symmetry. These three classes appear in shape transformations as $\overline{m}_0$ is decreased, starting from the SS class. The E1-10/11/12 shapes are effectively discocytes for which a 10/11/12-fold periodic undulation of the rim has broken the AD axisymmetry. The sketchy data suggest that the trend in the series of discontinuous transformations from E1-12 to E1-9 is a decrease in the number of bulges on the rim with decreasing $\overline{m}_0$. The interrelations between the AD and E1-n phases are complex. It appears that the E1-n sheets bifurcate separately from the AD sheet, as we shall discuss further in Section 2.7.5.

K and KE1-A: Knizocytes

Classically the term knizocyte (Bessis 1973) simply designates a triconcave red-cell shape. Such shapes can have two-fold or three-fold symmetry. They

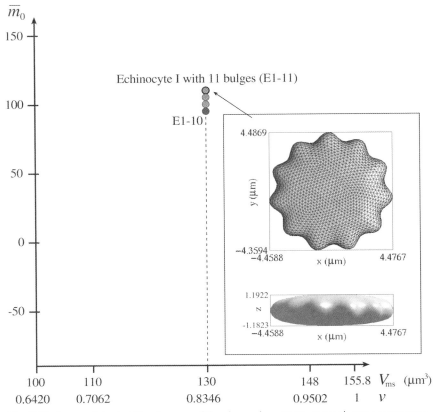

Fig. 2.36 Incomplete stability diagram of the shape class E1-11. Inset shows a representative shape at $V_{ms} = 130\ \mu m^3$ and $\overline{m}_0 = 110$.

appear in our calculations either as smooth shapes (K) or with varying degrees of echinocyte-I-type rippling of the margins (KE1). In our terminology, K and KE1-A are superclasses containing both two-fold shapes with a single mirror plane and shapes with a three-fold rotation axis. The stability diagrams of the K and KE1-A shape classes are shown in Fig. 2.29. The representative shapes K(1) and K(2) of the K class are shown in Figs. 2.38 and 2.39. The K(1) shape, at higher $\overline{m}_0$, appears to have three-fold symmetry. The K class changes from monoconcave at low $\overline{m}_0$, for example, K(2), to triconcave at high $\overline{m}_0$, for example, K(1). Note the resemblance of these shapes to the experimentally observed knizocyte shown in Fig. 2.3(C). The representative shape KE1-A(1) of KE1-A is shown in Fig. 2.40. Stability of these two classes requires a spherical or nearly spherical S_0. The K class transforms continuously into the KE1-A class at high $\overline{m}_0$ and continuously into the AS

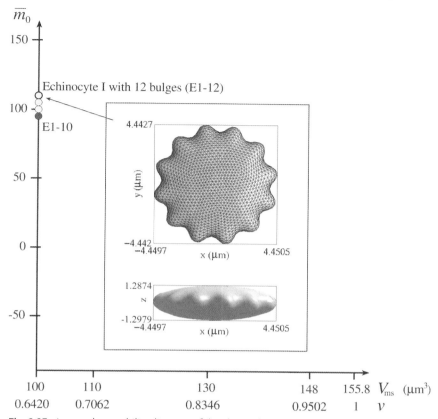

Fig. 2.37 Incomplete stability diagram of the shape class E1-12. Inset shows a representative shape at $V_{ms} = 100\ \mu m^3$ and $\overline{m}_0 = 110$.

at low $\overline{m}_0$. The KE1-A class transforms discontinuously into the KE1-B class at the upper KE1-A stability boundary.

KE1-B: Knizo-Echinocytes

The stability diagram and representative shapes KE1-B(1) and KE1-B(2) of the KE1-B shape class are shown in Fig. 2.41. KE1-B is initially mirror-symmetric with a flattened, ellipsoidal base. This mirror symmetry is lost as nine bulges develop on the rim of the base with increasing $\overline{m}_0$. Like K and KE1-A, KE1-B shapes require a spherical or nearly spherical S_0 in order to be locally stable. The constant surface feature of the KE1-B class is its pinch with three bulges. KE1-B transforms discontinuously into SS2 and the mirror-symmetric KE1-A upon crossing the upper and lower stability boundaries, respectively.

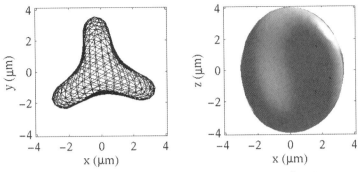

Fig. 2.38 Representative "knizocyte" shape K(1) at $V_{ms} = 155.8\ \mu m^3$ and $\overline{m}_0 = 80$. This shape is keyed to Figs. 2.24 and 2.29. A laboratory knizocyte image is shown as Fig. 2.3. Such shapes are observed for human red cells under certain conditions (see also Section 2.2.3).

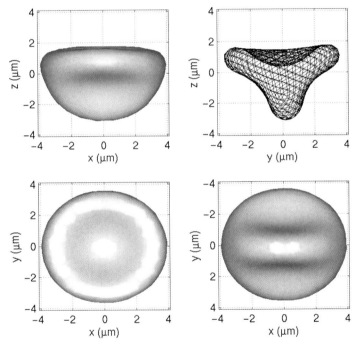

Fig. 2.39 Representative "knizocyte" shape K(2) at $V_{ms} = 155.8\ \mu m^3$ and $\overline{m}_0 = 60$. This shape is keyed to Figs. 2.24 and 2.29.

KE1-C: Knizo-Echinocytes

The incomplete stability diagram of KE1-C, together with a representative shape, is shown in Fig. 2.42. KE1-C shapes have mirror symmetry. Shapes with three-fold symmetry may also occur. KE1-C is another class that requires

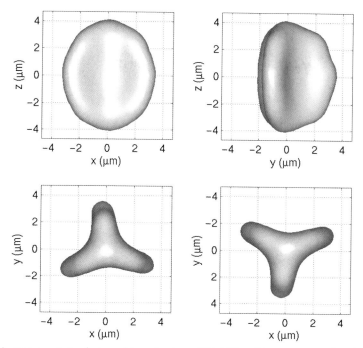

Fig. 2.40 Representative knizo-echinocyte shape KE1-A(1) at $V_{ms} = 155.8\ \mu m^3$ and $\overline{m}_0 = 95$. This shape is keyed to Figs. 2.24 and 2.29.

a spherical or nearly spherical S_0 in order to be locally stable. KE1-C shapes appear discontinuously as $\overline{m}_0$ is decreased, starting from the SS class. KE1-C is characterized by shapes with four bulges on each of their three edges plus mirror symmetry. KE1-C transforms discontinuously into KE1-B upon crossing the lower stability boundary.

SS2: Spiculated Shapes

The incomplete stability diagram of the SS2 class is shown in Fig. 2.43 together with a representative shape. The SS2 class may possess mirror symmetry. It is yet another special class that requires a spherical or nearly spherical S_0 in order to be locally stable. The SS2 shape is characterized by the emergence of a cluster of six spicules and a flattened base with nine bulges. Five of the six spicules are arranged pentagonally, with the remaining one located at the centre. SS2 transforms discontinuously into KE1-B at the lower stability boundary.

2 Red Blood Cell Shapes and Shape Transformations

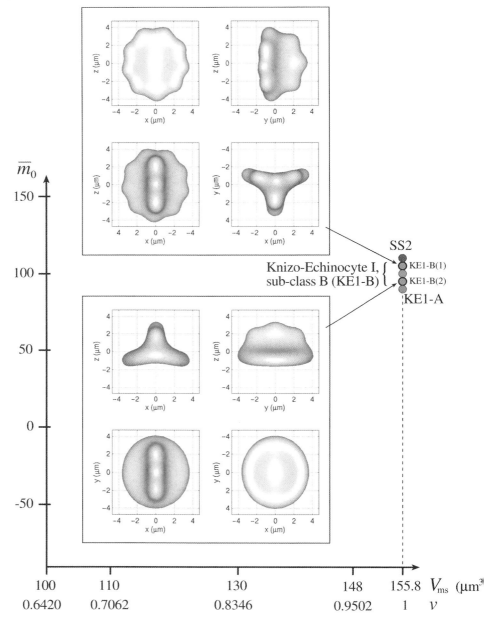

Fig. 2.41 Stability diagram of the KE1-B shape class. The insets show representative shapes KE1-B(1) and KE1-B(2) and key to Figs. 2.24 and 2.29.

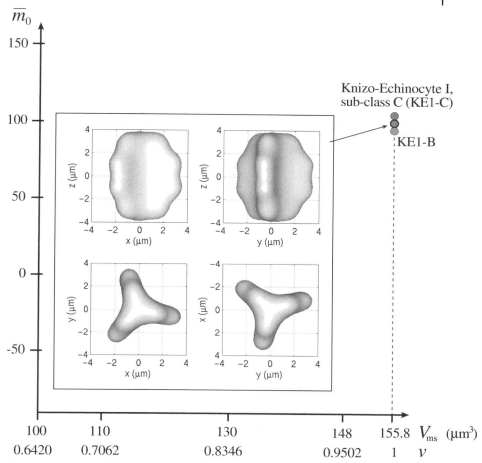

Fig. 2.42 Incomplete stability diagram of the shape class KE1-C. Inset shows a representative shape at $V_{ms} = 155.8\,\mu m^3$ and $\overline{m}_0 = 100$.

SS: Spiculated Shapes

The stability diagram and representative shapes SS(1) to SS(23) of the SS shape class are shown in Fig. 2.44. SS is a superclass defined to include a large collection of locally stable shapes with spicules and no obvious shape symmetry. As $\overline{m}_0$ increases, the spicules becomes shorter, sharper and more numerous, while the main body changes from a disc to an oval. This behavior mirrors what is observed experimentally during echinocytosis. Comparison between the shape sequences at different cytoskeletal reference volumes shows that V_{ms} (i.e., the shape of S_0) drastically affects the distribution of spicules on the main body. Noteworthy features include:

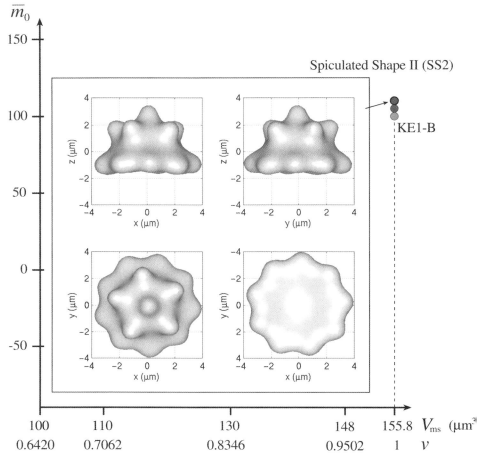

Fig. 2.43 Incomplete stability diagram of the shape class SS2. Inset shows a representative shape at $V_{ms} = 155.8\ \mu m^3$ and $\overline{m}_0 = 110$.

- Spicules tend to appear at locations on S where the corresponding locations on S_0 have a large positive mean curvature. Thus, when S_0 is relatively deflated (low V_{ms}), spicules increasingly tend to congregate at the rim of S, which corresponds to the part of S_0 where the mean curvature is most positive. Comparison of the shapes at $\overline{m}_0 = 140$, namely, SS(1), SS(5), SS(10), SS(13) and SS(17), shows that S_0 must be highly inflated for the spicules on S to be regularly spaced like those on an echinocyte III.

- When S_0 is spherical, the main body of the SS shapes is always oval, as it is for the classical echinocyte III class, but never disc-like, as it is for the classical echinocyte II class.

2.6 Predicted Shapes and Shape Transformations of the RBC

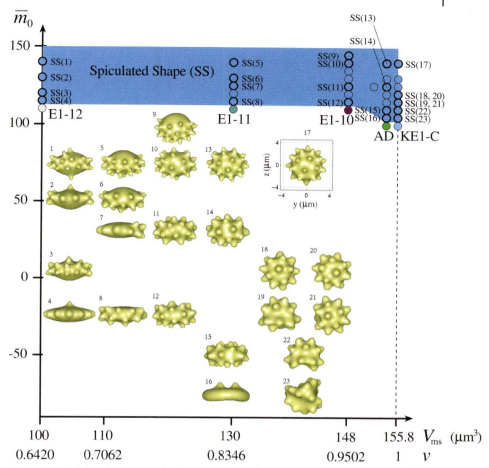

Fig. 2.44 Stability diagram of the SS shape class. The region of stability is shaded. The shape classes of selected points just outside the stability boundaries are indicated. Representative shapes SS(1) to SS(23) are shown below the stability region. The numbering is keyed both to the labeled points inside the stable region and to the corresponding points in phase-trajectory diagrams Figs. 2.18 to 2.24.

- If $V_{ms} < 154\,\mu m^3$, the SS class transforms discontinuously into an echinocyte I-like class (E1-10/11/12) at the lower stability boundary, as indicated in Fig. 2.44. Otherwise, the SS class transforms discontinuously into the AD or the KE1-C class. Therefore, $V_{ms} < 154\,\mu m^3$ is required for our predicted shape transformations to agree with the experimental observation that an echinocyte II transforms into an echinocyte I.

- We have not carried out a systematic investigation of the spicule distributions in the large SS superclass; nevertheless, it is interesting to note that several of the images in Fig. 2.44 show that spicules tend to congregate.

This behavior suggests that the interactions between spicule "excitations" of the surface are weakly attractive at long distances and repulsive at short distances.

Figure 2.45 summarizes the stability diagrams of all the major shape classes. It will play a key role in the following discussion.

2.7
Significant Results and Predictions

This section highlights the significant conclusions that can be drawn from the numerical membrane-mechanics results presented in Section 2.6. Our calculations have mainly (but not exclusively) been carried out with the parameter choices given in Table 2.2 and discussed in Appendix A. Our quantitative conclusions are, therefore, contingent on these values being correct. Some of these choices seem quite firm; however, others remain uncertain. In Section 2.8, we will comment further on the extent to which our main conclusions are robust against uncertainties in parameter values.

2.7.1
Observed SDE Shape Classes all Occur

The occurrence of the traditionally observed SDE shape classes is a necessary condition for validation of the model. Given the many different shape classes that have been observed, the many sheets of locally stable shapes and the dramatic sensitivity of shapes to small changes in control parameters, the broad agreement of observed and predicted shapes is encouraging.

The fact that this agreement is obtained with "reasonable" parameter values is also notable. In particular, it follows from Eqs. (2.23) and (2.33) that

$$\overline{m}_0 = \frac{2\pi R_A}{D_0}\left(\frac{\Delta A_0}{A_0} + \frac{D_0 C_0}{\pi \alpha_b}\right), \quad (2.101)$$

so that a change in the fractional area difference $\Delta A_0/A_0$ of 1% produces a shift in $\overline{m}_0$ of 105, that is, the whole range of the vertical axis in Fig. 2.45 corresponds to a fractional area change of about 2.5%, a number which is in the range suggested by experiments, as we shall discuss further in Section 2.8.5. Note that the sensitivity of RBC shapes to what appear to be small changes in the fractional area difference arises because of the large ratio R_A/D_0 in Eq. (2.101) and the fact that, in dimensionless form, it is $\overline{m}_0$ (or equivalently $\overline{C}_0$, Eq. (2.33)) which sets the curvature scale.

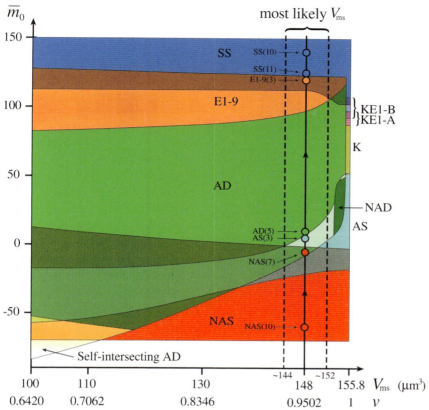

Fig. 2.45 Combined stability diagrams of the NAS, AS, AD, NAD, E1-9, K, KE1-A, KE1-B and SS shape classes. Only inside the "most likely" region, $144\,\mu m^3 < V_{ms} < 152\,\mu m^3$, does the sequence of predicted shapes agree with the observed SDE sequence of shape transformations. The ascending trajectory (i.e., increasing values of $\overline{m}_0$) at $V_{ms} = 148\,\mu m^3$ includes the shapes NAS(10), NAS(7), AS(3), AD(5), E1-9(3), SS(11) and SS(10). These calculated shapes are compared with laboratory images of the SDE sequence in Fig. 2.3.

2.7.2
Reference Shape S_0 of the Membrane Skeleton is an Oblate Spheroid

Figure 2.45 shows the combined stability diagram of the classes NAS, AS, AD, NAD, E1-9, K, KE1-A, KE1-B and SS for relaxed membrane-skeleton volumes between $V_{ms} = 100\,\mu m^3$ (the discocyte) and and $V_{ms} = 155.8\,\mu m^3$ (the sphere). Examination of this plot shows that the observed SDE sequence, illustrated in Figs. 2.2 and 2.3, is predicted to occur only in a narrow range of volumes centered at $V_{ms} \approx 148\,\mu m^3$ ($v = 0.950$). The lower limit of this range, at $V_{ms} \approx 144\,\mu m^3$ ($v = 0.925$), is the intersection point of the upper stability boundaries of NAS and AS; the upper limit, at $V_{ms} \approx 152\,\mu m^3$ ($v = 0.976$), is the intersection point of the lower stability boundaries of

E1-9 and SS. The correspondence of observed and predicted shapes within this range is illustrated in Fig. 2.3, where laboratory images of shapes from the SDE sequence are compared with the seven representative predicted shapes NAS(10), NAS(7), AS(3), AD(5), E1-9(3), SS(11) and SS(10) from the sequence of shape transformations at $V_{\mathrm{ms}} = 148\,\mathrm{\mu m}^3$, obtained by increasing $\overline{m}_0$ from -60 to $+140$ (see Fig. 2.21). The unstressed shape S_0 with $V_{\mathrm{ms}} = 148\,\mathrm{\mu m}^3$ is shown in Fig. 2.7.

How sensitive is the width of this allowed region to the poorly determined nonlinear coefficients a_3, a_4, b_1 and b_2 in Eq. (2.28)? The lower limit is quite insensitive, because, at the point where the upper stability boundaries of the NAS and AS classes intersect, the membrane skeletons of corresponding shapes in the two class are only weakly strained and remain in the linear elastic regime. On the other hand, the upper limit is in principle dependent on the nonlinear terms, since shapes in the SS class do experience large nonlinear strains. Nevertheless, this dependence is not expected to be strong, since this limit is set by the lower stability boundary of SS, where the corresponding shapes are the least strained of the class. Any change in the other parameters of F will, of course, shift the positions of both limits.

Various authors (Zarda et al. 1977; Evans and Skalak 1980; Fischer et al. 1981; McMillan et al. 1986; Li et al. 2005) have speculated that the unstressed shape of the membrane skeleton may be either a sphere ($v = 1$) or, on the other hand, a replica of the discocytic shape of the RBC ($v \approx 0.6$); however, these speculations do not seem to be based on strong evidence. Our results suggest that both these limiting cases are unlikely. If the unstressed shape were discocytic, we predict from our results at $V_{\mathrm{ms}} = 100\,\mathrm{\mu m}^3$ (Fig. 2.18) that the stomatocyte I stage would not be part of the main sequence of shape transformations and that there would not be spiculated RBC shapes resembling echinocytes II and III. While these predictions depend on the shape of S_0, they do not depend on a_3, a_4, b_1 and b_2. If the unstressed shape were a perfect sphere, we predict from our results at $V_{\mathrm{ms}} = 155.8\,\mathrm{\mu m}^3$ (Fig. 2.24) that the discocyte would not occur in the main sequence of shape transformations and that there would not be spiculated RBC shapes resembling echinocytes I and II. These predicted shape disappearances are the direct result of S_0 becoming a sphere and do not depend on a_3, a_4, b_1 and b_2. Our results suggest that the most likely unstressed shape of the membrane skeleton is an oblate spheroid (a sphere flattened at the poles) with a reduced volume in the range $0.925 \leq v \leq 0.976$. Such a shape is not unreasonable. It is consistent with recent experiments by Fischer (2004) which show that the erythrocyte retains "shape memory" after tank treading. It is known that the immature RBC (reticulocyte) has a highly irregular, folded shape (Bessis 1973). We expect, therefore, that the unstressed shape of its membrane skeleton is highly irregular. As the RBC matures and enters the circulatory system, the cumulative effect of the incessant deformation it experiences most likely molds the unstressed shape of its membrane skeleton into a regular, roughly spherical shape. That a discocyte is unstable when the unstressed shape is a sphere has been predicted independently Li et al. (2005); however, their results must be taken with some reservation, given the flaw in the numerical method they used to approximate the bending energy of the plasma membrane (see Section 2.3.5).

2.7.3
Predicted Hysteresis and Fluctuation Effects in RBC Shape Transformations

It is clear from phase-trajectory information (Figs. 2.18 to 2.24) and the composite stability diagram (Fig. 2.45) that over large regions of the phase diagram several different shapes are locally stable simultaneously. Furthermore, as we have seen, the energy barriers separating these local energy minima are frequently (but not always) large on the scale of $k_B T_{\text{room}}$, so excited-state branches are often metastable on experimental timescales. In this situation the boundaries between shape classes on specific experimental trajectories through the phase diagram – and, indeed, even the sequence of shape classes seen – may be history dependent. In addition, in regions near continuous shape transitions and in other regions where the shape-energy landscape is flat (on the scale of $k_B T_{\text{room}}$), observable shape fluctuations are predicted. Thus, the model makes testable predictions about hysteresis and fluctuations along specific trajectories.

Consider in this connection the SDE sequence of shape transformations predicted by the model in starting at low $\overline{m}_0$ in the stomatocyte range and increasing $\overline{m}_0$ through the discocyte and into the echinocyte range versus starting in the echinocyte range and decreasing $\overline{m}_0$ through the EDS sequence. For the sake of specificity we choose $V_{\text{ms}} = 148\,\mu\text{m}^3$ and consider variation of $\overline{m}_0$ only, as would be produced by slowly adding to the solution one of the echinocytogenic or stomatocytogenic agents listed in Table 2.1. Predictions for such a cyclic trajectory can be read off from Fig. 2.21. Assume that the starting shape is in the NAS superclass at $\overline{m}_0 = -60$. The predicted stable shape here is a curved invagination with a single mirror plane (NAS(10) in Fig. 2.27). As $\overline{m}_0$ is increased, this shape will transform most likely continuously into one with two mirror planes (NAS(9)) and on to the axisymmetric or nearly axisymmetric shapes NAS(8) and NAS(7). Then, near $\overline{m}_0 = 0$, we predict a discontinuous (abrupt) transition to the AS class followed shortly by another discontinuous transition to the axisymmetric discocyte class AD, which has a wide region of stability at this value of V_{ms}. The transition out of the AD class and into E1-9 at $\overline{m}_0 \approx 100$ is predicted to be continuous and, therefore, likely to involve appreciable shape fluctuation in the transition region (more in Section 2.7.5). Finally, around $\overline{m}_0 \approx 115$, we predict a discontinuous transition to the spiculated SS superclass of shapes, with a central body which is at first elliptical but becomes progressively more spherical as $\overline{m}_0$ increases. This sequence of transformations is illustrated in Fig. 2.3, with a schematic representation of the energy sheets. By contrast, the reverse trajectory is expected to be somewhat different. We predict that, upon decreasing $\overline{m}_0$, the shape remains in the SS superclass until $\overline{m}_0 \approx 105$ and then goes through E1-10 before reaching E1-9 at or close to the continuous transition back to to the discocyte AD. The energy landscape is rather flat in this region (see Section 2.8.5) and the E1-n sheets are close together, so the sequence may be masked by thermal fluctuations, although the hysteresis in the SS class should be clear. Further decrease of $\overline{m}_0$ brings about a discontinuous transition to the axisymmetric stomatocyte at $\overline{m}_0 \approx -10$ followed by another discontinuous transition to NAS at $\overline{m}_0 \approx -25$ (note the significant hysteresis at both these transition boundaries).

The upshot is that, overall, the same major shape sequence is expected in the ascending and descending trajectories. This is consistent with existing observations. Nevertheless, there are differences in the predicted shape classes (the appearance of E1-10 in the descending trajectory only[35]) and quantitative differences in the locations of some transition boundaries.

The sequence of these shape transitions was fixed by our choice of the relaxed cytoskeletal volume V_{ms}; however, the type of transition (continuous or discontinuous) and the corresponding predictions of hysteresis in the phase boundaries and specific regions of large fluctuation are new. To the best of our knowledge, no systematic experimental work on these issues has yet appeared in the published literature. We hope that our studies will stimulate such work. Numerical comparison of predicted transition boundaries with experiment will require finding ways of quantitating the changes in $\overline{m}_0$, that is, the changes in the area difference ΔA_0 of the plasma membrane (assuming, of course, that C_0 remains fixed). We will comment on this in Section 2.8.5. Nevertheless, such qualitative features as the type of transition and the presence or absence of observable fluctuations should not be hard to see.

In private correspondence, Fischer (2006) has reported careful qualitative observations of RBC shape sequences of the type described here. The level of agreement with our predictions is good but not complete. The principal discrepancies are in details of the stomatocyte transitions. If these differences hold up, they may signal the necessity of further fine-tuning of our parameter choices.

35) In fact, as we will see in Section 2.7.5, there is a narrow region where the true ground state is E1-10 but the energy barriers are sufficiently low so that at T_{room} the shapes present as a fluctuating thermal mixture of E1-9 and E1-10.

On a more general note, we end this section by pointing out that shape multiplicity – the coexistence of several different mechanically stable shapes under the same conditions – and related hysteresis effects have not been significantly discussed in the RBC shape literature. We hope that we have demonstrated the importance and usefulness of considering these issues. Indeed, there are suggestions in the literature that hysteretic effects involving branches outside the main SDE sequence may have been observed. A pertinent experimental example that has received scant attention is the occasional transformation of a stomatocyte II to a triangular stomatocyte instead of a stomatocyte III (Bessis 1972), which indicates that the triangular stomatocyte and the stomatocyte III have overlapping regions of stability. Indeed, the theoretical counterpart to the triangular stomatocyte, NAS-3 of Fig. 2.25, is found at the same V_{ms} and $\overline{m}_0$ as the theoretical counterpart to the stomatocyte III, NAS(10) of Fig. 2.27.

2.7.4
Strain Distribution over the Membrane Skeleton

In addition to determining the shapes that minimize F, our data allow us to calculate the distribution of the area strain α and the shear strain β over each shape. We show in Fig. 2.46 the predicted strain fields α and β for the seven calculated shapes of Fig. 2.3 at $V_{\mathrm{ms}} = 148\,\mu\mathrm{m}^3$. Stomatocytic shapes are typically highly dilated at the centre of the invagination and highly sheared and compressed near the rim of the invagination. The rest of the surface is only weakly strained. Spiculated shapes, on the other hand, typically have large dilation at the top of each spicule, large shear strain in the spicule neck region and significant compression in the main body.

The qualitative behavior of the membrane skeletal strain distribution over an echinocyte III shape can be visualized without resorting to equations. The presence of a more-or-less uniform distribution of similar spicules over the echinocyte body means that one can understand the strain distribution by looking at a single representative spicule, as illustrated in Fig. 2.47. The initially flat, unstressed patch of membrane skeleton is represented by the equally-spaced concentric rings at the left. When this patch is deformed at constant overall area to fit over the spicule (Fig 2.47), a process which is driven by the positive value of $\overline{C}_0$ in the bending energy, the central region (7) must stretch and the outer regions (1 + 2) must contract to provide the area for the out-of-plane deformation, both without significant shear. By contrast, in the intermediate regions (3–6) the dominant deformation is shear, accompanied by some stretch towards the apex (6) and some compression towards the base (3). The expected mem-

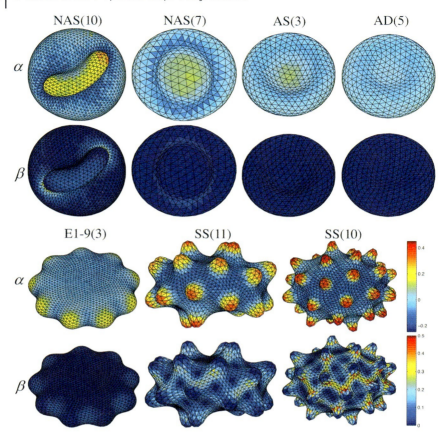

Fig. 2.46 Distributions of the strain fields α (stretch) and β (shear) over representative shapes belonging to some of the major shape classes. These shapes all have $V_{ms} = 148\ \mu m^3$ and correspond to the ascending sequence marked in Fig. 2.45 and illustrated in Fig. 2.3 a-g. Note the relatively low strain values for the non-spiculated shapes. As a consequence, these shapes are relatively insensitive to the nonlinear elasticity, Eq. (2.28). Spicules are characterized by large positive α at the tips, large β along the sides, and large negative α at the base and are, therefore, sensitive to nonlinear elastic parameters (see Section 2.8.3).

brane skeletal strain in the vicinity of the spicule apex bears a striking qualitative similarity to the observed membrane skeletal strain in the vicinity of the tip of a RBC tongue aspirated into a micropipette (Lee et al. 1999).

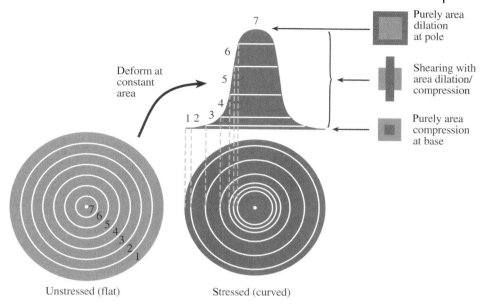

Fig. 2.47 Expected membrane-skeletal strain in the neighborhood of a spicule of an echinocyte III. When the flat unstressed patch at the left is deformed to fit over the spicule shown above the central region (corresponding to the tip of the spicule) is expanded, the outer region (corresponding to the base) is contracted and the middle region (corresponding to the sides) is sheared.

2.7.5
Large Thermal Fluctuations at the AD-to-E1 Boundaries

Our calculations suggest that, when the normal discocyte is pushed towards the echinocytic side, the first shape transition is into one of the E1-n shape classes, characterized by the development of relatively weak regularly-spaced undulations around the disc boundary. These shapes have been shown in Fig. 2.3 and in Figs. 2.32 to 2.37. In particular, at our preferred cytoskeletal volume $V_{ms} = 148 \, \mu m^3$, we might expect (see Fig. 2.21) the AD shapes to transform to E1-9 via a continuous (ct) transition before developing spicules to enter the SS class (dct) and similarly on a descending trajectory only with a small region of E1-10. In the lab, however, this detailed sequence is not seen. Indeed, as Fig. 2.3(e) suggests, none of the clearly periodic E1-n structures are identifiable. Instead, the initial smooth axisymmetric discocyte seems to go through a series of slowly-fluctuating shapes, characterized by "lumpy" irregularities, mostly but not entirely confined to the region of the disc boundary and on a scale which could roughly be 1/9 or 1/10 of the red-cell circumference, before suddenly developing the well-defined, fixed spicules characteristic of the SS phase. We argue in this section that this apparent discrep-

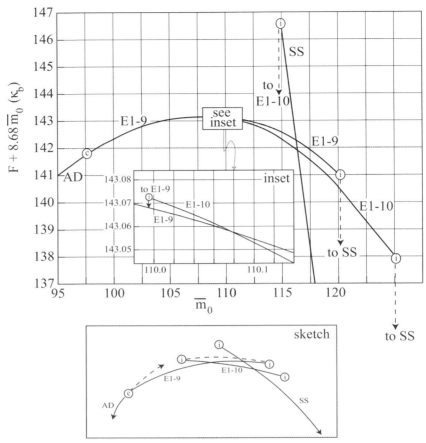

Fig. 2.48 Energy branches AD, E1 and SS at $V_{ms} = 148\,\mu\mathrm{m}^3$. The notations c and i denote a continuous transition and an instability boundary, respectively. Vertical dotted arrows indicate the shape class following the instability. The inset gives the local structure near the E1-9-to-E1-10 boundary (note the very narrow range in both energy and $\overline{m}_0$). The sketch below shows the probable interrelations of the AD, E1-9 and E1-10 branches. The dotted lines indicate stationary branches which are local saddle points and, therefore, do not show up in Monte Carlo studies: At the left is the unstable continuation of the AD branch; at the right is a "wing structure" of the type shown in Fig. 2.17.

ancy probably reflects the effects of normal thermal fluctuations and is, in fact, a validation of the model.

Figure 2.48 shows a plot of the energies of the various shape branches as a function of $\overline{m}_0$ through this region. Note that a term linear in $\overline{m}_0$ has been added to the energy in order to make the topology of the plot easier to see. Such a term does not affect the vertical offsets. The key point is the scale of the energy differences between the various branches through this region, as illustrated most clearly in the inset.

Note that at $\overline{m}_0 = 110.00$, where the descending E1-10 reverts discontinuously to E1-9, the offset of the energy levels is about $0.004\,\kappa_b$. This run was made at a computational temperature of $k_B T_{\text{cmpt}} \sim 0.001\,\kappa_b$, so the fact that the transition occurred at the next step shows that the energy barrier is also on this scale. By contrast, the thermal energy scale $k_B T_{\text{room}} \sim 0.02\,\kappa_b$ is larger by a factor of about $20\times$. Clearly, at $\overline{m}_0 = 110.00$, the room-temperature ensemble is a thermal mixture of E1-9 and E1-10 and a range of other nearby shapes without well-defined symmetries. By the same logic, we would not expect any sharp E1-9 to E1-10 or E1-10 to E1-9 transitions at room temperature as $\overline{m}_0$ is slowly changed, only a graded cross-over. Because our Monte Carlo method can only find the energy minima and does not generally allow us to probe the energy maximum between them, it is not immediately clear over how large a range of $\overline{m}_0$ this ensemble behavior is predicted to persist. However, some insight is possible if we make the working assumption that the scale of local variation of the energy surface is given roughly by the offset of the E1-9 and E1-10 energies. Thus, at $\overline{m}_0 = 115$ the offset is $0.25\,\kappa_b$, only a factor of 10 larger than the room-temperature thermal energy, suggesting that thermal fluctuations are likely important at least in the range $\overline{m}_0 = 105$–115.

So far we have discussed only the E1-9 and E1-10 branches; however, it is probable that the AD branch is also close in energy over much of this region. The transition between AD and E1-9, which takes place in the range $\overline{m}_0 = 95$–100, is predicted to be continuous. Near this transition, a simple Landau-theory analysis (see below) suggests that, at the transition, the AD branch switches character from a local minimum to a saddle and the whole energy-sheet structure is very flat over a range of $\overline{m}_0$ above and below the $T = 0$ transition point. Indeed, if we adopt the above estimate for the width of the region over which the energy surfaces are flat (i.e., have energy variation less than, say, $10\,\kappa_b T_{\text{room}}$), then it is quite possible that thermal fluctuations may be relevant at room temperature over a range between $\overline{m}_0 = 90$ and $\overline{m}_0 = 115$, where the spiculated SS shapes first come in.

It is possible to construct a Landau model for the energies in this region of the form

$$F(x,y) = F_0 + \frac{1}{2}r_1 x^2 + \frac{1}{4}u_1 x^4 + \frac{1}{2}r_2 y^2 + \frac{1}{4}u_2 y^4 + sx^2 y^2, \quad (2.102)$$

where x and y are order parameters associated with the E1-9 and E-10 phases, respectively, and $r_1 \sim (\overline{m}_1 - \overline{m}_0)$, $r_2 \sim (\overline{m}_2 - \overline{m}_0)$, with $\overline{m}_1 < \overline{m}_2$. At low values of $\overline{m}_0$, where r_1 and r_2 are both positive, the symmetric (AD) state with $x = y = 0$ has the lowest energy. However, as $\overline{m}_0$ increases above $\overline{m}_1$, there is a continuous transition to a state (E1-9) with $x \neq 0, y = 0$ and then later (for appropriately chosen parameters) to a state with $x = 0, y \neq 0$ via a discontinuous transition. The topology of this sequence of phase transitions is shown in the lower sketch of Fig. 2.48, where the dotted lines show energies of unstable stationary points. Such a model is at least consistent with the Monte Carlo results.

2.8
Discussion and Conclusions: The Future

The key ingredients of this work include the formulation of a mechanical model based on the composite structure of the red-cell membrane and the specification of the energetics of this model by means of an energy functional, $F_\mathrm{m}[S_0; S]$, of the membrane shape S and the shape S_0 of the relaxed membrane skeleton. We have argued that this functional depends on a small number of material moduli, most of which can be inferred from independent experiments. Finally, we have described a numerical method for finding red-cell configurations which are mechanically stable, that is, which correspond to local minima of the energy functional F_m at fixed area and volume.

We have found – perhaps somewhat surprisingly – that, despite the simplicity of the model, there is large collection of distinct minimum-energy surfaces corresponding to different shape classes, which overlap and interleave with one another, cross one another and bifurcate from one another as the principal control parameters of the system are continuously varied. Each of these sheets has its own region of stability and, because the distinct energy minima are often (but not always) separated by energy barriers significantly larger than the thermal energy $k_B T_\mathrm{room}$, there are many regions of the phase (shape) diagram over which a stable, ground-state shape can coexist for experimentally long times with one or more shapes belonging to higher-energy, metastable sheets. Knowing the basic geometry of these sheets and the magnitudes of the energy barriers which separate them, we can then make predictions about the specific sequence of shapes that a red cell will pass through as the control parameters of the system are changed on timescales which are long compared to hydrodynamic relaxation times but still short compared to the times required for full thermodynamic relaxation. This overall structure seems likely to be robust, even if future experiments should require some modification of the form of the energy functional or of the material parameters.

This concluding section is devoted to discussion of the significance and limitations of the kind of shape-mechanics calculations described above.

2.8.1
Validation of the Bilayer–Couple Hypothesis

In a seminal paper Sheetz and Singer (1974) proposed what they dubbed the "bilayer–couple hypothesis" as an explanation for the SDE transformations. It was known at the time that the erythrocyte plasma membrane was asymmetrical in the sense that the compositions of the inner and outer leaflet were different. Sheetz and Singer proposed in a purely qualitative way that various amphipathic compounds bind preferentially into the outer leaflet ("crenators," that is, what we have called echinocytogenic agents in Table 2.1) or the inner leaflet ("cup-formers," that is, what we have called stomatocytogenic agents), thus causing a bending moment tending to promote evagination or invagination, respectively. In our terminology this is equivalent to the identification of the area difference ΔA_0 as the primary driving force in the SDE transformations.[36] The original proposal was formulated in connection with a pure bending model of the plasma membrane and made no mention of the membrane skeleton. We now know that without the membrane skeleton the region in phase space of high $\overline{m}_0$ would be dominated by budding and vesiculated shapes (Waugh 1996; Iglič 1997; Wortis 1998) rather than echinocytes, so that, in its original form, the bilayer–couple mechanism was incomplete. On the other hand, the insight that the bending tendency, expressed quantitatively as the parameter $\overline{C}_0$ or $\overline{m}_0$ in our model, is the basic driving force behind the SDE transformations is, in our view, correct. Thus, in the broader sense, our study may be regarded as a final validation of the original Sheetz-Singer mechanism.

Although this mechanism was proposed more than 30 years ago, until now there has not been a quantitative demonstration that the observed shapes and shape transformations can, indeed, arise in detail from the bilayer–couple mechanism. The main reasons for this 30-year hiatus were, initially, the lack of a specific mechanical model for the shape mechanics and, when that was available, the lack of sufficient computational power for testing it. Realistic model testing, as we have done in this study, involves intensive calculations of shapes that need not be axisymmetric (such as echinocytes). Computers commonly available before the late 1990s were not powerful enough to perform such a task in a reasonable time. Our ability to explore quantitatively the implications of a specific mechanical model has made possible a remarkable level of validation of the bilayer–couple mechanism. The fact that the shapes and shape transformations of the SDE sequence seen in experiments do appear in the membrane-mechanics calculations in a detailed and highly

36) Equivalently, one can imagine inducing the bending moment by substituting big-headed (or small-tailed) amphiphiles in the outer leaflet to increase C_0 with no change of leaflet area (or the reverse in the inner leaflet to decrease C_0), since ΔA_0 and C_0 enter together in the driving parameter $\overline{m}_0$ (or $\overline{C}_0$).

non-trivial way provides very strong evidence for the validity of the bilayer–couple mechanism as the driving force for shape change.

Further testing of the bilayer–couple mechanism and the mechanical model will need to focus on quantifying the link between the concentration of each echinocytogenic (or stomatocytogenic) agent and specific changes in the mechanical moduli. We hope and believe that detailed tests of this type will be forthcoming and that they will pin down further the values of the parameters that govern the shape mechanics. We believe that the interplay between theory and experiment must play a central role in this effort. In particular, it will be important to have a computer program that is able to predict shapes based on given mechanical-parameter inputs. We hope that we have supplied a model for such a program.

2.8.2
Generalized Phase Diagrams and Trajectories

In this presentation we have focused on the effective spontaneous curvature $\overline{C}_0$ of the plasma membrane (as encoded in the dimensionless variable $\overline{m}_0$) as the primary driving force in the SDE transformations and on the relaxed volume V_{ms} as an important (but unknown) parameter necessary to characterize the membrane skeleton. This reduces the effective phase space to two dimensions. The further assumption that addition of shape-changing agents influences only $\overline{C}_0$ means that all experimental "trajectories" are along the vertical direction in our standard $(\overline{m}_0, V_{\mathrm{ms}})$ phase space. These assumptions have provided convenient restrictions on the range of parameter values over which we have needed to make computations; however, they are certainly oversimplifications.

Any parameters appearing in the energy functional $F_{\mathrm{m}}[S_0; S]$ can in principle be regarded as variable, thus providing additional dimensions to a generalized phase space and corresponding generalized phase and stability diagrams. The poorly determined values of the skeletal elastic moduli μ and K_α are good examples, as are the nonlinear coefficients a_3, a_4, b_1, b_2, etc. (Similarly but less relevantly, perhaps, more parameters describing the shape of S_0 could be introduced.) Although we have not explored such additional dimensions in any systematic manner, some comments on sensitivity to the material parameters regarded as fixed in Table 2.2 are provided in the next section. If and when the predictions (Sections 2.6 and 2.7) of our particular model are found to be in disagreement with experiment, it will become important to examine a range of the less-well-determined material parameters.

In addition, it is far from clear that shape-changing chemical agents act exclusively on $\overline{m}_0$. Although driving via $\overline{m}_0$ may well be the dominant effect, it is likely that these agents have some concentration-dependent influence on the material moduli and even on the cytoskeletal parameters. If such influences are present, then slowly changing the concentration of one chemical agent drives the system along a trajectory in the generalized phase space which is still one-dimensional but no longer along a single phase-space axis. Thus, the analysis of shape-change sequences along the lines of Section 2.7.3 would require the appropriately generalized stability diagrams.

2.8.3
Sensitivity of Results to Variation of Elastic Parameters

We have emphasized in Section 2.3.4 that many variables enter the shape problem in dimensionless, scale-invariant combinations. Any parameter changes which affect these combinations will modify the predicted RBC shapes and shape transitions. Such potential changes are relevant, since several of the parameters we have listed in Table 2.2 are not well determined by independent experiments, as discussed in Appendix A. We focus in this section on the skeletal elasticities, as these appear to be among the more-poorly determined parameters.

2.8.3.1 Effects of Varying μ

We have tested the effect of varying the shear modulus μ while preserving the relation $K_\alpha = 2\mu$ (see Appendix A). We find that increasing μ from $2.5\,\mu\mathrm{J\,m}^{-2}$ to $6\text{–}9\,\mu\mathrm{J\,m}^{-2}$ (the range of estimates from micropipette aspiration experiments) causes the region of stability of the NAD class to spread to lower values of V_ms. Of course, the boundaries of all other shapes classes also shift. Thus, the AS $\rightarrow$ AD shape-class transformation with increasing $\overline{m}_0$ in the range $144\,\mu\mathrm{m}^2 \lesssim V_\mathrm{ms} \lesssim 152\,\mu\mathrm{m}^2$ (see Fig. 2.3) is replaced by an AS $\rightarrow$ NAD $\rightarrow$ AD shape-class sequence, which is inconsistent with observation. On the other hand, halving the value of μ to $1.25\,\mu\mathrm{J\,m}^{-2}$ causes the disappearance of shapes resembling echinocytes I, II and III. In place of echinocytic shapes, shapes with several long arms appear as $\overline{m}_0$ increases. In particular, the AD $\rightarrow$ E1-9 shape-class transformation at $\mu = 2.5\,\mu\mathrm{J\,m}^{-2}$ is replaced by a transformation from the AD class to a class of starfish-like shapes not seen in experiments.

2.8.3.2 Higher-Order Nonlinear Elastic Terms

A crucial part of our model is the inclusion of higher-order nonlinear elastic terms in the Taylor expansion of $f_\mathrm{ms}(\alpha, \beta)$, Eq. (2.26), in order to harden the membrane-skeletal elasticity at moderate-to-large values of the strains α and β. As discussed in Appendix A, the coefficients a_3, a_4, b_1 and b_2 of these additional terms are not measured directly but have been fitted in a rather crude manner to a high-deformation approximation. It may prove possible to extract the values of these parameters by using techniques based on optical tweezers (Hénon et al. 1999; Sleep et al. 1999; Dao et al. 2003; Lim et al. 2004a; Lim et al. 2004b; Mills et al. 2004; Li et al. 2005) or the optical stretcher (Guck et al. 2001); however, this has not yet been done, and in any case these additional terms affect only the finer details of highly strained shapes. Specifically, they affect the size and shape of spicules belonging to shapes in the SS superclass and the size, shape and orientation of the invaginations of shapes corresponding to the lower half of the NAS superclass.

We have tested the effects of eliminating these higher-order terms by setting a_3, a_4, b_1 and b_2 to zero. We find that the elimination of these terms has no drastic effect on shapes with surface features that are rather smooth: shapes resembling the stomatocytes III, II and I, discocytes and echinocytes I of the experimental SDE transformations continue to remain locally stable. This insensitivity to the higher-order elastic terms is not unexpected, since for smooth shapes (e.g., shapes (a) to (e) in Fig. 2.3) cytoskeletal deformations remain in or near the linear-elastic regime, as is clear from Fig. 2.46. On the other hand, the effect of eliminating the higher-order terms on shapes in the spiculated shape classes (e.g., Fig. 2.3(f) and (g)) is significant. In the absence of nonlinear hardening of the skeletal elasticity, the bending elasticity of the plasma membrane begins to dominate for these spiculated RBCs, thus forcing significant regions of the membrane skeleton to undergo large and rapidly-varying local stretch and shear deformations. Spiculated shapes exhibit large variations of f_{ms} in the large-deformation regime. If the higher-order terms are eliminated, thereby weakening the in-plane elasticities, we would expect the shape to exhibit bending-dominated behavior, characterized by the formation of small spherical buds joined via narrow necks to a larger body (see Fig. 2.5). We were not able to study directly the locally stable spiculated shapes resulting from elimination of the higher-order terms. The reason for this is that, without the hardening effect of the higher-order terms, our program finds (in this region of the stability diagram) only non-physical, numerical artifacts whose triangular surface elements fail the checks for shape regularity that are built into our program. This comes about because we are effectively constrained to a maximum of 5120 triangles in the computational mesh, which provides insufficient resolution to represent the small buds that would occur in the absence of the higher-order terms.

2.8.4
Understanding the Action of Shape-Change-Inducing Agents

Some important shape-change-inducing agents for the SDE transformations were listed in Table 2.1; a more extensive list is provided in Wong (1999). Many, if not all, of these effects can now be understood qualitatively, based on the simple bilayer–couple picture that $\overline{m}_0$ (or $\overline{C}_0$) is the principal driving force (Steck 1989). Thus, it follows from Eqs. (2.23) and (2.33) that agents which act to increase either the area difference ΔA_0 or the spontaneous curvature C_0 will be echinocytogenic, while those which act to decrease these quantities will be stomatocytogenic. For example, any compound added to the solution which partitions preferentially into the outer leaflet of the plasma membrane will tend to increase ΔA_0 and is expected to be echinocytogenic, while any compound which partitions to the inner leaflet will be stomatocytogenic. Similarly, any large-headgroup amphipath will have an echinocytogenic effect if it partitions to the outer leaflet (thus increasing C_0) and a stomatocytogenic effect if it partitions to the inner leaflet (thus decreasing C_0), and the reverse for amphipaths with large tails and/or small heads.

It is easy in this context to understand quantitatively the effects of many amphipathic drugs. Recall from Section 2.2.1 that the inner leaflet of the plasma membrane is rich in negatively charged lipids. It follows that positively charged (cationic) amphipaths will tend to partition preferentially into the inner leaflet, thus increasing its area, decreasing ΔA_0 and acting as stomatocytogenic agents, in agreement with Table 2.1. By contrast, negatively charged (anionic) amphipaths are expected to be echinocytogenic. Similarly, cholesterol is known to partition preferentially to the outer leaflet (Steck et al. 2002), so, when added to the solution, it acts echinocytogenically, as shown in Table 2.1. Finally, hypertonic salt is presumed to screen the negative charges of the inner leaflet, thus decreasing its area relative to that of the outer leaflet and acting echinocytogenically.

Some effects are less easy to explain. It is known, for example, that the lipid asymmetry of the leaflets is maintained by ATP-dependent flipases. Thus, if the red cell is deprived of ATP, passive lipid redistribution between the leaflets is expected, leading to a change of ΔA_0. However, the sign of this effect – whether it should be echinocytogenic (as it is) or the reverse – is not clear without further understanding of the specific mechanisms. The pH effect also remains open. One simple hypothesis would be that pH titrates lipid charge groups. If this were applied to the negatively charged inner-leaflet membranes, we would expect acidic pHs to neutralize the inner leaflet, thus decreasing its area. But this effect would be echinocytogenic, quite the opposite of what is observed. Indeed, work by Gedde et al. (1995) and Gedde et al. (1997) has shown that lipid asymmetry and the presence of inner-leaflet titratable groups do not have any strong connection to the pH effect. On the other hand, earlier works by Elgsaeter et al. (1986), Stokke et al. (1986a) and Stokke et al. (1986b) have shown that the membrane skeleton expands in vitro in response to high cytoplasmic pH, suggesting that the effect may be associated with the membrane skeleton rather than the plasma membrane. A possible mechanism for this has been proposed by Mukhopadhyay et al. (2002). Recent work (Gedde et al. 1999) suggests that competing effects are at work. The full mechanism of the pH effect has not yet been definitively established. Glass surfaces are known to modify the local pH, so it is presumed that the glass effect and the pH effect are related.

2.8.5
Experimental Quantitation of $\overline{m}_0$

Our model predicts a specific, quantitative correspondence between each shape and a corresponding value of $\overline{m}_0$. Thus, for any observed shape we can infer a corresponding value of $\overline{m}_0$. In order to test such a model, it is necessary to have an independent measurement of $\overline{m}_0$ or, equivalently, to have a second mechanical prediction – beyond the erythrocyte shape – which can be compared with experimental results.

While this program has not been carried out for our model, a similar test has been carried out recently by Kuzman et al. (2004). These authors performed micropipette aspiration on a series of strongly echinocytic cells. By measuring the length of the membrane projection pulled into the pipette as a function of aspiration pressure and fitting to a mechanical model very similar to ours, they inferred $150 \leq \overline{m}_0 \leq 225$.[37] The lower end of this range is in rough agreement with our value of $\overline{m}_0 = 140$ for the echinocyte III-like SS(10) shape of Fig. 2.44. This agreement should probably be regarded as somewhat fortuitous, since Kuzman et al. (2004) used bending and elastic moduli somewhat different from ours[38] and solved their model in an approximate manner using a parametrized spicule shape.

Several older studies (Lange and Slayton 1982; Ferrell et al. 1985; Chi and Wu 1990) estimated the relative area difference between the two leaflets of the plasma membrane. These estimates are, at best, semi-quantitative. Furthermore, the quantity that was estimated in these studies is the actual relative area difference $\Delta A/A_0$ rather than the initial or preferred relative area

---

37) The quantity denoted $\overline{\Delta a_0}$ in Kuzman et al. (2004) is equivalent to our $\overline{m}_0/4\pi$. The fit is shown in Fig. 8 of the cited reference.

38) The parameters that differ in values from ours are the volume V_{RBC}, the shear modulus μ, the stretching modulus $K_\alpha = 2\mu$ and the non-local bending modulus $\overline{\kappa}$. They used $V_{RBC} = 109\ \mu m^3$, $\mu = 6\ \mu Jm^{-2}$ and $\overline{\kappa} = (8 \times 10^{-19}\ J)/\pi$, whereas we used $V_{RBC} = 100\ \mu m^3$, $\mu = 2.5\ \mu Jm^{-2}$ and $\overline{\kappa} = (4 \times 10^{-19}\ J)/\pi$. Their use of large values of μ and K_α but without our nonlinear terms may be regarded as a way of effectively capturing the elastic hardening at large deformation, as is relevant for the spicules. How effective this form may be at capturing the variation of f_{ms} over the full echinocyte III membrane skeleton is not clear.

difference $\Delta A_0/A_0$. If the nonlocal bending modulus $\bar{\kappa}$ were large, these two quantities would be the same; however, because $\alpha_b \sim 1$, the most one may hope for is that they should be similar in scale. Specifically, Lange and Slayton (1982) estimated that the actual relative area difference required for echinocytosis is about 1%. Ferrell et al. (1985) estimated that the actual relative area difference required to induce stage III echinocytes is $(1.7 \pm 0.6)\%$. Chi and Wu (1990) estimated that the actual relative area difference required to induce stage III echinocytes is $(3.2 \pm 0.2)\%$. All these estimates are in the same general range and suggest that area differences of the order of a few percent are required for echinocytosis. Our value of $\bar{m}_0 = 140$ for the SS(10) shape, which corresponds via Eq. (2.101) to $\Delta A_0/A_0 = 1.3\%$, is entirely consistent with this range.

Ideally, of course, one would like to be able to "dial" $\bar{m}_0$ systematically by making controlled changes in the chemical environment. Comprehensive reviews of the biochemical effects of various chemical agents on the plasma membrane may be found in Deuticke et al. (1990), Schreier et al. (2000) and Boon and Smith (2002). However despite some qualitative success, as discussed in the previous section, quantitative control over $\bar{m}_0$ in the lab has not been achieved. To do so would require a quantitative, predictive understanding of the chemical and physical determinants of the area difference ΔA_0 and the spontaneous curvature C_0, which is not yet available. Thus, as of this writing, it is not possible to prepare a red cell with a known $\bar{m}_0$, to observe its shape and to compare with theory.

In practice what is done is simply to add slowly to the extracellular solution measured quantities of some shape-change inducing agent and to watch over time as the shape evolves. Interpreting such experiments is complicated by the different timescales that influence the observations. Consider the effects of exogenous amphipathic drugs, whose qualitative action we think we understand (see Section 2.8.4). When such agents are added to the exterior solution, they partition first into the outer leaflet of the plasma membrane and then, by a flip-flop process, into the inner leaflet, from which they equilibrate finally with the interior solution. The eventual equilibrium distribution of the drug is controlled by the relative affinity of the additive for the different leaflets; however, the time to reach equilibrium and the state of the system during relaxation depend on the rates for each of the transport processes. Thus, if the flip-flop is slow, the drug may build up in the outer leaflet during the initial part of the process, causing a transient echinocytic effect, even if the equilibrium distribution favors the inner leaflet and produces stomatocytosis at long-enough times (Isomaa et a. 1987). In general, the processes which control interleaflet lipid (and drug) transport are both active and passive. In the normal resting state the asymmetric lipid distribution of the RBC is maintained by ATP-dependent phospholipid translocases (Alberts et al. 2002) which appear to respond on scales of 10–20 minutes. Passive transport, which can be driven both by diffusive gradients and by membrane tension, induces shape change on one-minute time scales. The time scale for the hydrodynamic processes of purely mechanical shape equilibration are a few seconds. Other relevant time scales include the time for mixing the drug in the solution and the time for partitioning between the solution and the adjacent leaflet. Assume for simplicity that the mixing and partitioning times are fast and consider the sequence of events after drug addition. For times shorter than one second, cell shapes are controlled by membrane dynamics. For times between one second and one minute, one sees a transient sequence of shapes corresponding to a changing $\Delta A_0(t)$, as the drug equilibrates passively between leaflets. For times between one and ten minutes, one may hope to observe the equilibrium shape corresponding to the passively equilibrated ΔA_0. This would be the relevant range for confirming the equilibrium-shape predictions, if it were possible (which it is not!) to calculate $\overline{m}_0$ from the drug concentration in solution. Finally, on time scales longer the 10 minutes, the lipid translocases would start to have an important effect on $\overline{m}_0$. Analogous dynamical effects have been observed in pipette aspiration experiments (Raphael and Waugh 1966; Artmann et al. 1997; Svetina et al. 1998; Raphael et al. 2001; Kuzman et al. 2004).

2.8.6
Effects of Lateral Inhomogeneity of the Red-Cell Membrane

The membrane-mechanics model we have used assumes spatial homogeneity of the plasma membrane and the membrane skeleton. This approximation breaks down at sufficiently short length scales. As long as the scale of inhomogeneity remains small on the scale of variation of the shape topography, coarse-graining arguments suggest that homogeneity should remain a good approximation. Among the shapes we have treated, it is the echinocytes which have the sharpest spatial features, with spicules on the scale of several tenths of microns.

Inhomogeneities in the composition of the plasma membrane may be of two types: those that form spontaneously by active or passive processes unrelated to membrane shape and those that occur in response to membrane stresses such as local curvature. Of the former type are the so-called "rafts" which have received much attention in the recent literature (Simons and Ikonen 1997; Jacobson et al. 2007). Most, but not all, of these are on scales smaller than $0.1\ \mu\mathrm{m}$ and need not concern us here, although it would be interesting to know whether they congregate preferentially at spicules or stomatocytic invaginations. An example of the latter type would be the migration of large-head lipids to regions of significant positive curvature such as the apex region of spicules. This tendency of molecules to segregate to relieve steric stresses must, of course, be balanced against the entropically-driven tendency towards uniform mixing. There is some experimental evidence for effects of both kinds. First, Rodgers and Glaser (1991), Rodgers and Glaser (1993a), Rodgers and Glaser (1993b) and Welti and Glaser (1994) have observed regions of inhomogeneity larger in size than $0.1\,\mu m$ in the membranes of rabbit RBCs. They found that domains enriched in either phosphatidylcholine or phosphatidylserine form in intact membranes not subjected to an inducing agent, suggesting that these lipid domains occur naturally. In addition, when they exposed intact membranes to CPZ, a fluorescent and stomatocytogenic compound, they observed the formation of CPZ-enriched domains. Rodgers and Glaser osmotically swelled the CPZ-laced RBCs to spheres because they needed simple cell geometries for fluorescence imaging. Thus, it is not known whether or not the CPZ-enriched domains form preferentially in the invaginations of the stomatocytes. Secondly, a recent study by Baba et al. (2004) of echinocytosis of human RBCs induced by poly (ethylene glycol)-cholesterol (PEG-Chol) found that the distribution of PEG-Chol in the plasma membrane changes from uniform at the echinocyte I stage to non-uniform at the echinocyte II stage. They found that PEG-Chol is located preferentially at the spicules of an echinocyte II shape, that is, in the regions with the largest positive curvature. It remains to be seen whether all echinocytogenic chemical compounds have such effects.

The results of Rodgers and Glaser and Baba et al. suggest that the model of the plasma membrane may eventually need to take into account the effects of coupling between curvature and membrane composition. This improvement would not be difficult to accomplish theoretically; however, such a model would introduce additional unknown parameters and does not appear to be necessary to account for the main features of the SDE sequence.

The membrane skeleton maintains its fixed connectivity on the time scales of shape observations, so that it is not subject to segregation effects like the plasma membrane; however, its intrinsic inhomogeneity is on the scale of the cytoskeletal spectrin mesh (see Section 2.2.2), which is on the order of $0.1\,\mu m$. Thus, at the smallest spicule sizes we are treating, the approximation of continuum elasticity is already at or close to its limit. At shorter scales such as those occurring near sphero-echinocytosis (Section 2.2.3), it will certainly break down. Thus, in our opinion, it is likely that the vesiculation which is observed at sphero-echnocytosis and sphero-stomatocytosis is the result of a budding of the plasma membrane between the cytoskeletal anchor proteins when the inverse of the effective curvature C_0^{eff} reaches the mesh size.

2.8.7
Membrane Mechanics of RBCs of Other Mammals

The structure and behavior of red cells belonging to other mammalian species are often (but not always) similar to those of humans. Similarities include the composite quasi-two-dimensional structural organization of the membrane, the discocytic shape under normal physiological conditions (there are some exceptions like camels and llamas) and the gross shapes (differences do exist in the finer surface features) and shape transformations (spiculation or cupping) induced by echinocytogenic or stomatocytogenic agents (Jain and Kono 1972; Jain and Keeton 1974; Jain 1975; Smith et al. 1980; Smith et al. 1982). These similarities should allow the model proposed in this study to be used in analyzing the biomechanics of other mammalian RBCs, provided that the mechanical parameters specific to each species are known. The shear moduli for a few mammals other than human have been estimated. They include the rabbit, the rat, the opossum and the llama (whose RBCs are not discocytic) (Waugh and Evans 1976; Waugh 1992). These moduli were obtained using the micropipette aspiration technique and under the assumption of local area incompressibility of the membrane skeleton. Micropipette aspiration probes the nonlinear in-plane elasticities at high deformation (see Appendix A). Therefore, the measured shear moduli for the aforementioned mammals should not be equated to the linear modulus μ of Eq. (2.27). The RBC area and volume for the rabbit, mouse, rat and hamster have also been measured (Waugh 1992). There are currently to our knowledge no measurements of the other mechanical parameters (see Table 2.2) for mammals other than humans. It is hoped that this work will stimulate experiments to probe in detail the membrane mechanics of other mammalian RBCs. However, at present the ingredients for detailed testing of our approach on other mammalian species do not appear to be available.

2.8.8
Summary

In summary, we have proposed a continuum elastic model of the RBC that is the most general and realistic to date. The essential features of this model are (i) the inclusion of the plasma-membrane bending elasticity and the membrane skeletal stretch and shear elasticities, (ii) the inclusion of higher-order nonlinear elastic terms in the Taylor expansion of the membrane-skeletal strain-energy density f_{ms} and, on the calculational side, (iii) an effective numerical method for representing simultaneously the membrane shape and cytoskeletal conformation in three dimensions without any prior assumption of shape symmetry.

2.8 Discussion and Conclusions: The Future

The total free energy F of our model contains two unknown control parameters, $\overline{m}_0$ (or, equivalently, $\overline{C}_0$) and V_{ms}. The former governs the bending tendency of the plasma membrane; the latter characterizes the unstressed shape of the membrane skeleton. We assume that an observed RBC shape is a locally stable shape that minimizes F at given values of $\overline{m}_0$ and V_{ms}. In practice, the minimization of F can only be performed numerically. We have chosen to minimize F over a computational mesh using a Monte Carlo technique. By performing minimizations systematically over a range of increasing and decreasing values of $\overline{m}_0$ at fixed V_{ms}, we have mapped out the ranges of stability of a number of distinct shape classes. We have been able to deduce from this sheet structure the predicted sequence of shapes and shape transformations as $\overline{m}_0$ is changed by echinocytogenic and/or stomatocytogenic agents. By matching this sequence to the observed SDE transformations, we have inferred that the unstressed membrane-skeletal shape of the RBC is likely to be an oblate spheroid with a volume in the range $144\,\mu\mathrm{m}^3 \lesssim V_{\mathrm{ms}} \lesssim 152\,\mu\mathrm{m}^3$ (i.e., a reduced volume in the range $0.925 \lesssim v \lesssim 0.976$). The fact that it is possible to make this match provides a detailed validation of the Sheetz–Singer bilayer–couple hypothesis. Our results show further that hysteresis is generally to be expected in any cycle of shape transformations, and we are able to make specific predictions. Finally, we have made the first comprehensive predictions of the area and shear strain fields over every RBC shape of the SDE sequence.

These predictions are partly generic, and in this respect we believe that they will prove robust. However, they are dependent in detail on the values of material moduli some of which are poorly determined. We will not be surprised if further experiments lead to some fine-tuning of the moduli. This process will likely involve a careful interplay of theory and experiment. On the one hand, this model makes specific predictions which may or may not be falsified by experiment. On the other hand, the measurement of mechanical moduli involves the observation of shape change, so the interpretation of experiments will increasingly require theoretical input of the type illustrated here.

We hope that the analytical framework developed here will both provoke and make possible future experiments. A start in this direction has already been made by Khairy et al. (2007), who have used urea to weaken the red-cell membrane skeleton and have analyzed the resulting morphological changes by corresponding shape calculations of the type described here.

In closing we wish to acknowledge the assistance and encouragement we have received for this project over the last two decades from Myer Bloom, David Boal, Evan Evans, Tom Lubensky, Narla Monads, Erich Sackmann, Ted Steck and other participants in the now-terminated program of the Canadian

Institute for Advanced Research entitled "The Science of Soft Surfaces and Interfaces."

Appendix A
Material Parameters and Related Experiments

In this appendix we collect material pertaining to the numerical values of important parameters which enter the cell-shape calculations. The choices made for these parameters are crucial in that the results (for example, the phase diagrams) are complex and relatively small changes in (some) parameters can have profound effects. The final choices for these parameters are summarized in Table 2.2. As will become evident below, many of these parameters have been determined by different methods at different times with levels of agreement which are less than fully satisfactory. We will try to clarify, as far as we are able, the reasons for the disagreements and our rationale for choosing the values shown in Table 2.2. We have tried to make clear in the text the level of robustness of each of our principal results to changes in input parameters. Finally, we note that these are biological materials and a certain level of variability is entirely normal.

A.1
Geometry: Cell Area and Volume

Table 2.4 gives a sampling of area and volume measurements over the last 40 years with associated references and some notes. Observe that the natural variability is about $\pm 10\%$. We have chosen in Table 2.2 the representative values $V_0 = V_{RBC} = 100\,\mu\text{m}^3$ and $A_0 = A_{RBC} = 140\,\mu\text{m}^2$, which leads to $R_A = 3.34\,\mu\text{m}$ from Eq. (2.1). Because of the scaling discussed in Section 2.3.4, V_0 enters the shape problem only via the (dimensionless) reduced volume $v = 0.642$, as defined by Eq. (2.3).

A.2
Plasma Membrane Moduli

The bending moduli κ_b and $\bar{\kappa}$ of the RBC plasma membrane have been estimated experimentally. The experimental techniques used to extract κ_b include flicker spectroscopy (Brochard and Lennon 1975; Zilker et al. 1987; Zeman et al. 1990; Peterson et al. 1992b; Zilker et al. 1992; Strey et al. 1995; Humpert and Baumann 2003), micropipette aspiration of flaccid RBCs (Evans 1983), tether formation from the RBC membrane (Waugh and Bauserman 1995; Hwang and Waugh 1997) and local pulling of the RBC membrane using the tip of an atomic force microscope (Scheffer et al.

Table 2.4 A chronologically-ordered compilation of some previous measurements of area and volume for unfractionated adult blood samples in isotonic solutions at room temperature (20–25 °C) and pH = 7.4. The sample size is given in the form: number of subjects from whom blood was drawn / combined total number of cells. Each entry in columns 3 and 4 is given as: mean ($\pm$) standard error of mean ($\pm$) population standard deviation.

Reference	Sample size	Area (μm^2)			Volume (μm^3)		
		Mean	SEM	SD	Mean	SEM	SD
Canham and Burton (1968)[a]	7 /1016	138.1		17.4	107.5		16.8
Evans and Fung (1972)[b]	1 / 50	135		16	94		14
Jay (1975)[c, d]	7 / 788	136.9	0.5		104.2	0.6	
Jay (1975)[c, e]	7 / 843	133.4	0.5		98.1	0.6	
Fung et al. (1981)[b]	14 /1581	129.95	0.40[g]	15.86	97.91	0.41[g]	16.16
Linderkamp and Meiselman (1982)[f]	5 / 200	134.1		13.8	89.8		12.7
Nash and Meiselman (1983)[f]	/ 160	137	5.5		99	5	
Linderkamp et al. (1983)[f]	10 / 400	134.3	6.1	13.5	88.4	3.8	12.8
Linderkamp et al. (1986)[f]	5 / 150	132.1	6.7	14.1			
Linderkamp et al. (1986)[f]	10 / 300				94.9	5.0	13.7
Stadler and Linderkamp (1989)[f]	10 / 400	137.1	6.7	14.8	90.5	4.4	13.2
Waugh et al. (1992)[f]	1 / 65	135		10	93		12
Linderkamp et al. (1993)[f]	10 / 300	137.1	6.7		90.5	4.4	
Engström and Löfvenberg (1998)[h]	10 / 500	141.4	3.0		105.7	3.3	
Ruef and Linderkamp (1999)[f]	10 / 400	132		9	101		11

[a] Manual tracing of diametrical cross-sections of cells hanging vertically from the underside of microscope coverslips photographed edge-on.
[b] Determination of geometries of cells by interference holography.
[c] Digital tracing of diametrical cross-sections of cells hanging vertically from the underside of microscope coverslips photographed edge-on.
[d] In Ringer solution without albumin.
[e] In Ringer solution with albumin.
[f] Video recordings of geometries of micropipette-aspirated cells.
[g] From the text of Fung et al. (1981); values given in the abstract of Fung et al. (1981) are different (1.03 μm^2 for A and 1.06 μm^3 for V).
[h] Digitized images of diametrical cross-sections of cells hanging vertically from the underside of microscope coverslips viewed edge-on.

2001). The resultant estimates of κ_b are listed in Table 2.5. Except for the flicker spectroscopy estimates, the estimates of κ_b shown in Table 2.5 are in good agreement. The differences may be due to the different theoretical assumptions in the analyses of these experiments. In addition, it has been proposed by Strey et al. (1995) that the smaller estimates given by flicker spectroscopy of RBC membrane fluctuations at short wavelengths (Zilker et al. 1987; Zilker et al. 1992) may be related to active (ATP-driven) motion of the RBC membrane (Tuvia et al. 1998). Cuvelier et al. (2005) recently hinted at the possibility of estimating the κ_b of the plasma membrane using yet another technique, based upon the coalescence of two tethers pulled from the RBC membrane. Currently $\bar{\kappa}$ has been estimated only once, in a study of tether formation from the RBC membrane by Hwang and Waugh (1997), who obtained $\bar{\kappa} = (3.8 \times 10^{-19}\,\text{J})/\pi$. The values of κ_b and $\bar{\kappa}$ quoted in Table 2.2 are based on studies of tether formation from the RBC membrane (Waugh and Bauserman 1995; Hwang and Waugh 1997), which use the full ADE model Eq. (2.22) in the analysis. Finally, the Gaussian modulus κ_g has not been measured. The topological character of this term (Section 2.3.2) makes it difficult to approach. One possibility would be to find a way of comparing the energies of topologically different shapes, for example, before and after a fission event. Another would be to find a way of measuring the local boundary torque density, Eq. (2.70).

A.3
Membrane-Skeleton Moduli

A.3.1 Linear Moduli μ and $\mathbf{K}_\alpha$

Considerable uncertainty still attends the values of the linear moduli K_α and μ. Older estimates for these moduli based on micropipette aspiration of of intact RBCs (see Hochmuth (1993) for a review) give typical numbers in the range $\mu \sim 6\text{–}9\,\mu\text{J}/\text{m}^2$, appreciably larger than the value $\mu = 2.5\,\mu\text{J}/\text{m}^2$ we have adopted in Table 2.2. These estimates are based on fitting experimental data to the uncorrected, weak-deformation form Eq. (2.27). Furthermore, they are often based on analysis done under the assumption that the membrane skeleton is locally incompressible, $K_\alpha = \infty$, so $\alpha = 0$ everywhere.[39] Both of these assumptions are of doubtful validity.

Recent experiments by Discher et al. (1994), Discher and Mohandas (1996) and Lee et al. (1999) have conclusively disproved the assumption of local area incompressibility of the membrane skeleton. In addition, these authors found that λ_1 and λ_2 in the aspirated region of the RBC actually depart sig-

---

[39] The attractive feature of assuming local incompressibility is that it allows μ to be determined without the need to measure the variation of the principal extension ratios λ_i over the aspirated region of the RBC.

Table 2.5 Experimental estimates of the bending modulus κ_b of the human red blood cell plasma membrane.

Reference	κ_b ($\times 10^{-19}$ J)	Method
Brochard and Lennon (1975)	0.13 – 0.3	Flicker spectroscopic analysis of (1) correlation functions for thickness fluctuations at two different points;
Zilker et al. (1987)	0.34 ± 0.08	(2) Fourier modes (with wavelengths 0.5–1.0 µm) of RBC membrane deformation amplitudes in the normal direction;
Zeman et al. (1990)	≤ 2–3	(3) long-time decay of the autocorrelation function of thickness;
Zilker et al. (1992)	0.23 ± 0.05	(4) Fourier modes (with wavelengths 0.25–3 µm) of RBC membrane deformation amplitudes in the normal direction;
Peterson et al. (1992b)	1.4 / 4.3	(5) the thickness fluctuation profile along a diameter;
Strey et al. (1995)	2 – 7	(6) the first three azimuthal Fourier modes of fluctuations of the rim (wavelengths comparable to cellular dimensions); and
Humpert and Baumann (2003)	1.0 – 1.9	(7) the power spectra and autocorrelation functions of fluctuations of the cell center and cell rim.
Evans (1983)	≤ 1.8	Micropipette aspiration of flaccid RBCs.
Waugh and Bauserman (1995)	2.0 – 3.0	Tether formation from the RBC membrane: (1) pulling tethers from RBC membranes at constant force;
	1.8 – 2.7	(2) relaxation of tethering force at constant tether length;
Hwang and Waugh (1997)	2.0 ± 0.6	(3) relaxation of tethering force at constant tether length; and
	2.0	(4) pulling tethers from RBC membranes at constant velocity.
Scheffer et al. (2001)	2.07 ± 0.32	Pulling RBC membranes locally using an atomic force microscope.

nificantly from unity. The small-deformation regime applies for $\lambda_i = 1 + \delta_i$ with $|\delta_i| \ll 1$, in which case $\alpha^2 \approx (\delta_1 + \delta_2)^2$ and $\beta \approx (\delta_1 - \delta_2)^2/2$ and there is no ambiguity about the definitions of the moduli. The problem arises because in the micropipette experiments (and, also, for the more-extreme echinocyte and stomatocyte shapes) there are regions of the red cell where the δ_i's are not small, so terms of order δ^3 and higher may become important. The terms α^2 and β in Eq. (2.27) constitute a particular choice of such higher-order terms; however, there is no fundamental reason to expect nature to conform to this choice. That is, of course, the reason we have adopted the more-general expression Eq. (2.28). In summary, we regard these micropipette-based values of μ as unreliable both because they incorrectly assume incompressibility and because they infer a low-deformation modulus from a high-deformation experiment. If we are correct in thinking that the moduli harden at high deformation, it seems likely that inferring linear moduli from high-deformation experiments will over-estimate their values.

More recently, K_α and μ have been measured in the linear (or purportedly linear) regime by a different set of techniques, as summarized in Table 2.6. The estimates of Hénon et al. (1999), Guck et al. (2001) and Lee and Discher (2001) are based on intact RBCs, whereas the others are not. Note that there are no experimental estimates of K_α for intact RBCs. The estimates of K_α given by Lenormand et al. (2001) and Lenormand et al. (2003) are for bare membrane skeletons. It is not clear to what extent they may be regarded as applying to the intact RBC, since the physical and chemical conditions experienced by the skeleton in these two situations are quite different.

We first discuss estimates not based on intact RBCs. The analyses of Lenormand et al. (2001) and Lenormand et al. (2003) assume a homogeneous

Table 2.6 Recent estimates of K_α and μ using techniques other than micropipette aspiration.

Reference	K_α (μJ/m^2)	μ (μJ/m^2)	Method
Hénon et al. (1999)		2.5 ± 0.4	Pulling two beads attached to a discocytic or nearly spherical swollen RBC in diametrically opposite directions using optical tweezers.
Sleep et al. (1999)		200	Pulling two beads attached to a saponin-lysed spherical ghost in diametrically opposite directions using optical tweezers.
Guck et al. (2001)		13 ± 5	Applying optical stress fields using an optical stretcher to stretch osmotically swollen spherical RBCs.
Lee and Discher (2001)		1–10	Tracking thermal fluctuations of fluorescent beads 40 nm in diameter, attached to actin directly within ghosts or indirectly via glycophorin C outside RBCs.
Lenormand et al. (2001)	4.8 ± 2.7	2.4 ± 0.7	Pulling three beads attached to the periphery of a bare membrane skeleton in different directions using optical tweezers at an osmolality of 25 mOsm/kg.
Lenormand et al. (2003)	9.7 ± 3.4	5.7 ± 2.3	Same as previous but at an osmolality of 150 mOsm/kg.

linear elastic stress field over the triangular region formed by three beads attached to the periphery of a membrane skeleton. This assumption is questionable near the bead attachment points, where the deformation can be appreciable. How good the assumption of linear elasticity is depends critically on the area of the region affected by nonlinear elastic effects and the magnitude of $f_{\rm ms}$ in that region. If the nonlinear-elastic area is negligible compared to the linear-elastic one and if $f_{\rm ms}$ of the nonlinear-elastic area is not significantly above the linear elastic limit, then one can safely neglect nonlinear elastic effects. The value of μ quoted by Sleep et al. (1999) is much larger even than the micropipette-based values. We surmise that there is some systematic problem, although we can only speculate as to the cause. First, this measurement is based on RBC ghosts permeabilized by saponin, a process which may change the properties of the cytoskeletal proteins significantly. Furthermore, the analysis is based on the work of Parker and Winlove (1999), who used Eq. (2.27) with the unrealistic assumption of local area incompressibility of the membrane skeleton. Finally, the regions near the bead attachment points are highly stressed and likely to be in the large-deformation regime, where the effective in-plane elastic moduli harden significantly.

Next, we turn to the estimates based on intact RBCs. Lee and Discher (2001) find a broad but still reasonable range for μ through an analysis of the root-mean-square in-plane displacement of beads attached to the membrane skeleton. They assume that the bead displacements arise entirely from thermal fluctuations. It is not known whether or not ATP-dependent fluctuations (Tuvia et al. 1998) might be important for their analysis. Guck et al. (2001) produce small displacements by use of a so-called optical stretcher, a technique which avoids the point forces which occur in experiments based on optical tweezers and which complicate the analysis because of possible local nonlinearities. Unlike optical tweezers, the optical stretcher applies an optical stress field to the RBC membrane. This stress field is caused by the momentum transfer from two opposed unfocussed laser beams which are used to trap a RBC. Guck et al. (2001) used the optical stretcher to stretch osmotically swollen spherical RBCs into ellipsoids. Their analysis of the cell deformation gives a value for μ (purportedly in the linear elastic regime) that is somewhat higher than the range found in micropipette aspiration, which operates in the nonlinear elastic regime. This higher value may occur because of problems with their analysis: In treating the RBC membrane as a thin shell, they neglect the bending energy of the plasma membrane and do not separate the area compressibility of the plasma membrane from that of the membrane skeleton. These two area compressibilities operate at very different energy scales, as discussed in Section 2.3.5. In addition, they may have failed to take into account the important effects of F_V, the energy required to change the RBC volume from that set by the osmolarity of the suspending medium (see Sections 2.1.3 and 2.3.1). A spherical, turgid RBC has an elevated internal pressure that puts the RBC membrane under isotropic tension, according to the law of Laplace. In deforming the RBC shape from a sphere to an ellipsoid, the RBC volume will decrease. When the volume of a turgid RBC is forced to decrease there is a corresponding rise in the concentration of osmotically active molecules trapped inside the RBC. This increases the osmotic pressure difference across the RBC membrane according to the van't Hoff equation, Eq. (2.10), and, hence, increases the isotropic tension the RBC membrane is subjected to. Neglecting this rise in isotropic tension may artificially inflate the apparent value of μ to the higher value found by Guck et al. (2001). Finally, Hénon et al. (1999) investigate stretching of discocytic RBCs and nearly spherical RBCs using optical tweezers. Their value for μ, although consistent with our expectation, is also based on some questionable analysis. Specifically, they neglect the bending energy of the plasma membrane, assume local area incompressibility of the membrane skeleton, approximate the RBC membrane as two parallel independent discs with no stress at the edge in the case of a discocytic RBC and do not quantify the effects of point forces on the membrane skeletal deformation near the bead attachment points. All of these issues could affect the quoted results significantly.

In summary, we remain somewhat skeptical of all the new estimates of μ. Our simulation value $\mu = 2.5\,\mu\mathrm{J/m}^2$ (see Table 2.2) agrees with the work of Hénon et al. (1999) on intact RBCs and is consistent with Lee and Discher (2001) and Lenormand et al. (2001). We prefer a value lower than the micropipette-aspiration range because it gives the correct sequence of shapes in the SDE sequence (see Section 2.8.3).

This situation for the RBC membrane-skeleton stretch modulus K_α is unsatisfactory from an experimental point of view in that there are no measurements for intact cells. The measurements of Lenormand et al. (2001) and Lenormand et al. (2003) on isolated skeletons suggest that K_α is roughly a factor of two greater than μ. This relation $K_\alpha = 2\mu$ has also been predicted for models in which the membrane skeleton is approximated as a triangular network of springs (Hansen et al. 1996; Boal 2002). On this basis we have adopted the value $K_\alpha = 5\,\mu\mathrm{J/m}^2$ for our simulations.

A.3.2 Nonlinear Terms

Given the uncertainty surrounding the linear moduli, it is not surprising that the nonlinear parameters a_3, a_4, b_1 and b_2 which occur in Eq. (2.28) are poorly characterized. The basic issue is to reconcile the weak linear moduli of Hénon et al. (1999) and Lenormand et al. (2001), which are necessary to model the observed SDE sequence, with the higher values, $\mu \geq 6\,\mu\mathrm{J/m}^2$, which have been inferred from micropipette aspiration experiments at high

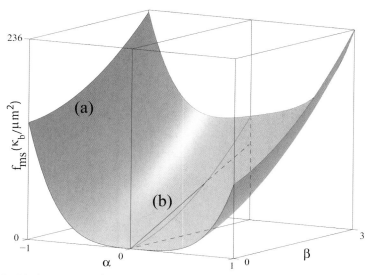

Fig. 2.49 (a) Elastic energy density $f_{\mathrm{ms}}(\alpha, \beta)$ of the membrane skeleton, Eq. (2.28), with the parameter choices of Table 2.2. The effect of the nonlinear terms is to harden the elasticity at high deformation. (b) Shown for comparison at $\alpha = 0$ is the straight line corresponding to a linear elasticity with the high-estimate value $\mu = 6\,\mu\mathrm{J/m^2}$ (see text).

deformation, as analyzed under the assumed constraint $\alpha = 0$ (incompressibility). Our point of view in selecting the parameters of F_{ms}, Eq. (2.28), is to adopt the low values of the linear moduli but to postulate nonlinear terms which "harden" the elasticity at higher deformation sufficiently to make it broadly compatible with the pipette-aspiration experiments (see below). Figure 2.49 shows a plot of our assumed elastic energy density, $f_{\mathrm{ms}}(\alpha, \beta)$, with $\mu = K_\alpha/2 = 2.5\,\mu\mathrm{J/m^2}$ and the nonlinear coefficients given in Table 2.2. It is relatively flat near the origin, due to the small linear moduli, but rises rapidly at larger strains. By way of comparison, Fig. 2.49 includes in the plane $\alpha = 0$ a plot of a purely linear elasticity $f_{\mathrm{ms}}(\alpha = 0, \beta) = \beta\mu$ with the higher, micropipette-based modulus, $\mu = 6\,\mu\mathrm{J/m^2}$. This line lies above the energy-density surface for $\beta < 1.9$ but below it thereafter. Our red-cell simulations sample a range from $|\alpha|, \beta \leq 0.1$ for the normal discocyte to $|\alpha|, \beta \sim 0.3$ for highly deformed shapes (see Section 2.7.4).

To represent the high-deformation regime we have adopted a form proposed by Discher et al. (1994) and Mohandas and Evans (1994) in connection with pipette-aspiration experiments,

$$f_{\mathrm{ms}} = \frac{K_N}{2}\left[(\lambda_1\lambda_2)^2 + \frac{2}{n(\lambda_1\lambda_2)^n}\right] + \frac{\mu_N}{2}\left(\lambda_1^2 + \lambda_2^2\right) \qquad (2.103)$$

or, equivalently, with a shift of origin so that $f_{\mathrm{ms}}(\alpha = 0, \beta = 0) = 0$,

$$f_{\mathrm{ms}} = \frac{K_{\mathrm{N}}}{2}\left[(\alpha+1)^2 + \frac{2}{n(\alpha+1)^n} - \left(1+\frac{2}{n}\right)\right] + \mu_{\mathrm{N}}\beta(\alpha+1), \quad (2.104)$$

where an irrelevant[40] term linear in α has been omitted. K_{N} and μ_{N} are, respectively, effective stretch and shear moduli at large deformation. The operational definition of a large deformation is $|\lambda_{1,2} - 1| \gtrsim 0.5$, that is, a uniaxial stretching or compression in a principal direction of at least 50%, as discussed in related work of Lee et al. (1999). Note that K_{N} and μ_{N} do not reduce to the linear elastic moduli in the limit $\lambda_i \to 1$. Discher et al. (1994) performed fluorescence imaging of the membrane skeletal deformation of the RBC projection aspirated into a micropipette. They determined experimentally that $K_{\mathrm{N}} \approx 2\mu_{\mathrm{N}}$ and $1 \leq n < 2$; however, they did not determine the actual values of K_{N} and μ_{N}. They argue that, effectively,

$$\mu \approx \frac{\mu_{\mathrm{N}} K_{\mathrm{N}}}{\mu_{\mathrm{N}} + K_{\mathrm{N}}}, \quad (2.105)$$

where μ is the older micropipette-based effective value of the linear shear modulus discussed in the previous section. Thus, to represent the high-deformation data, we have taken $n = 1$ along with $K_{\mathrm{N}}/2 = \mu_{\mathrm{N}} = 9\,\mu\mathrm{J/m}^2$ (corresponding to $\mu \approx 6\,\mu\mathrm{J/m}^2$).

In choosing the nonlinear parameters, a_3, a_4, b_1 and b_2, our aim is to construct Eq. (2.28) so that it interpolates between the weak linear elasticity discussed in the previous section and the strong effective elasticity described by Eq. (2.104). The values listed in Table 2.2 provide such an interpolation, as illustrated in Fig. 2.50, which compares (i) Eq. (2.104) with the parameters given in the previous paragraph; (ii) Eq. (2.28) with the parameter values shown in Table 2.2; and (iii) the purely linear elasticity model Eq. (2.27) with the weak linear moduli.[41] Different interpolations could easily be made, and they would provide energy densities somewhat different from ours, depending on the range of strains to be compared. Note, however, that the strains $|\alpha|, \beta \leq 0.3$, which are relevant to the SDE shapes, do not probe significantly into the high-deformation regime, so the effect of any such change on the predicted shapes would be expected to be small. In closing we stress that, from our point of view, it is not the specific values of the nonlinear parameters that are important but the general shape of the elastic-energy-density surface $f_{\mathrm{ms}}(\alpha, \beta)$. The weak linear elastic moduli are necessary to reproduce the proper SDE sequence. Changing the nonlinear parameters will only modify significantly the strongly echinocyte and stomatocyte shapes.

40) The integral of α over S_0 gives the total membrane area, $\int_{S_0} dA_0\, \alpha = \int_{S_0} dA = A_0$, which is fixed. Thus, any term linear in α simply adds a shape-independent constant to the total energy.
41) The shear part of Eq. (2.104) remains significantly stronger than the nonlinear shear we have used for $0 \leq \alpha, \beta \leq 1$, as shown in Fig. 2.50 (right). Nevertheless, the nonlinear shear increases as β^2 and always dominates at sufficiently large β, e.g., for $\alpha = 0$ the crossover occurs near $\beta = 3.47$.

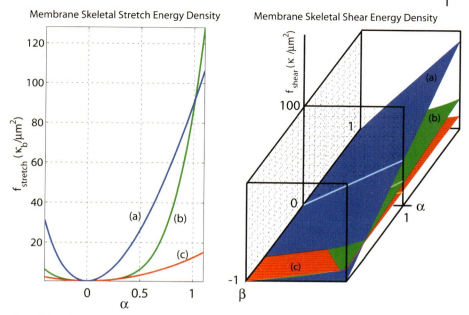

Fig. 2.50 Different representations of the stretch and shear parts of the elastic energy density. Left: (a) Stretch part of Eq. (2.104). (b) Fully nonlinear stretch, $f_{\text{stretch}} = \frac{K_\alpha}{2}\left(\alpha^2 + a_3\alpha^3 + a_4\alpha^4\right)$. (c) Linear stretch, $f_{\text{stretch}} = \frac{K_\alpha}{2}\alpha^2$. Right: (a) Shear part of Eq. (2.104). (b) Fully nonlinear shear, $f_{\text{shear}} = \mu\left(\beta + b_1\alpha\beta + b_2\beta^2\right)$. (c) Linear shear, $f_{\text{shear}} = \mu\beta$. Parameter values of Table 2.2 are used for (b) and (c). For (a), $K_N/2 = \mu_N = 9\,\mu\text{J}/\text{m}^2 = 45\,\kappa_b/\mu\text{m}^2$.

We hope that further experiments will in the future clarify the values of the linear elastic constants and the nonlinear parameters of the RBC membrane. As mentioned above, we believe that such experiments are best carried out *in situ* on intact red cells, for example, by techniques such as those of Guck et al. (2001) applied to flaccid RBCs. Insofar as such experiments infer the elastic parameters by observing shape deformations, the analysis will of necessity require a mechanical model of the membrane. We believe that the techniques described in this work will prove useful in this analysis.

Appendix B
Symmetry Sets the Form of Elastic Energies

This appendix aims to provide a brief but self-contained primer on the so-called "Landau" (symmetry) arguments that underlie the forms of the bending energies F_{sc} and F_{g}, Eqs. (2.14) and (2.15), and the stretch and shear energies F_{ms}, Eqs. (2.27) and (2.28).

B.1
Local Bending Energy

The argument starts at Eq. (2.19) and asks what form the local bending-energy density $f_b(\mathbf{r})$ can have that is consistent with (i) Euclidean invariance and (ii) in-plane membrane isotropy, which is a consequence of the assumption that the plasma membrane is in the isotropic fluid (L_α, smectic-A) phase.[42] There are, of course, many ways to run such an argument. What follows is one of the simplest.

The idea is to build f_b from quantities which characterize the local shape in a way which respects Euclidean invariance (the energy should not depend on where S is or how it is oriented) and membrane isotropy (one direction is as good as another in an isotropic 2D fluid). To parametrize S locally, construct the local tangent plane at $\mathbf{r}$ and let (x_1, x_2) be Cartesian coordinates in this base plane. The local shape of any smooth S can be represented as the perpendicular displacement h of S from the base plane (the "Monge representation"),

$$h(x_1, x_2) = \frac{1}{2} C_{ij} x_i x_j + \frac{1}{3!} D_{ijk} x_i x_j x_k + \ldots, \qquad (2.106)$$

with summation over repeated indices. The coefficients C_{ij}, D_{ijk}, and so on depend on $\mathbf{r}$ and parametrize the local shape of S in a way that respects Euclidean invariance, so that we may write schematically $f(\mathbf{r}) = f[\{C_{ij}(\mathbf{r})\}, \{D_{ijk}(\mathbf{r})\}, \ldots]$.[43] It is easy to verify that C_{ij} is just the curvature matrix (Appendix C). It can be made diagonal by a proper choice of the coordinate axes. The diagonal elements are just the principal curvatures C_i. Thus, with no loss of generality, we can write

$$h(x_1, x_2) = \frac{1}{2}\left(C_1 x_1^2 + C_2 x_2^2\right)$$
$$+ \frac{1}{3!}\left(D_1 x_1^3 + D_2 x_2^3 + D_3 x_1^2 x_2 + D_4 x_1 x_2^2\right) + \ldots, \qquad (2.107)$$

where now $C_1 \equiv C_{11}$, $D_1 \equiv D_{111}$, etc. Notice that, if the surface is smooth on some scale R (typically, for us, the scale R_A of the whole cell or that of smaller surface features such as spicules), then we expect that h/R is a function of x_1/R and x_2/R with coefficients of order unity. Thus, we expect $C_i \sim 1/R$, $D_i \sim 1/R^2$, etc.

The key question now is, "What combinations of the coefficients C_i, D_i, etc., can appear in f_b?" To approach this question, we organize the possibilities in powers of $1/R$. At zeroth order, there is always a shape-independent

[42] If the membrane were in a phase with additional in-plane order, e.g., smectic-C, then additional terms would appear.
[43] Note that higher-order coefficients like D encode information about the spatial derivatives of the curvature tensor $\mathbf{C}$ and are in this sense superfluous once $\mathbf{C}(s^1, s^2)$ is given.

constant, which plays no role in shape selection. At first order, there are only the terms in C_1 and C_2. Second-order invariants are C_1^2, C_2^2 and $C_1 C_2$ plus the four D_i, so

$$f_b[\{C_i\},\{D_i\},\ldots] = \kappa_0 + \kappa_1 C_1 + \kappa_2 C_2 + \kappa_3 C_1^2 + \kappa_4 C_2^2 + \kappa_5 C_1 C_2$$
$$+ \kappa_6 D_1 + \kappa_7 D_2 + \kappa_8 D_3 + \kappa_9 D_4 + \ldots, \quad (2.108)$$

where the coefficients κ_i are material moduli. Now an additional argument comes into play. If we want surfaces related by $x_1 \leftrightarrow x_2$ to have the same energy, then we must require that $\kappa_1 = \kappa_2$, $\kappa_3 = \kappa_4$, $\kappa_6 = \kappa_7$ and $\kappa_8 = \kappa_9$. Finally, the requirement that the reflected surfaces $x_1 \leftrightarrow -x_1$ and $x_2 \leftrightarrow -x_2$ have the same energy forces $\kappa_6 = \kappa_7 = \kappa_8 = \kappa_9 = 0$. The upshot is that through order $1/R^2$ there are only four independent terms, proportional to 1, $(C_1 + C_2)$, $(C_1^2 + C_2^2)$ and $C_1 C_2$.[44] After a redefinition of moduli, $\kappa_b \equiv 2\kappa_3, C_0 \equiv -\kappa_1/2\kappa_3, \kappa_g = \kappa_5 - 2\kappa_3$, these terms correspond precisely to the terms in $F_{sc} + F_g$. Note that we have not invoked $h \leftrightarrow -h$ as a symmetry (i.e., $C_i \leftrightarrow -C_i$, $D_i \leftrightarrow -D_i$, etc.), which effectively interchanges the inside and outside of the membrane. Such a symmetry, if it existed, would eliminate the energy term proportional to $(C_1 + C_2)$, that is, it would require $C_0 = 0$. Typically, however, for the plasma membrane $C_0 \neq 0$, reflecting the compositional asymmetry of the two leaflets noted in Section 2.2.

When the energy depends on the curvature only via the local invariants through second order, then stability places certain conditions on the moduli κ_b and κ_g. Thus, it is easy to show that,

$$\frac{A_1}{2}\left(C_1^2 + C_2^2\right) + A_2 C_1 C_2 + A_3 (C_1 + C_2) + \text{constant}$$
$$= \frac{A_1}{2}(C_1 + C_2 - C_0)^2 + (A_2 - A_1) C_1 C_2 + \text{constant}$$
$$= \frac{(A_1 + A_2)}{4}(C_1 + C_2 - C_0^*)^2 + \frac{(A_1 - A_2)}{4}(C_1 - C_2)^2 + \text{constant}.$$
$$(2.109)$$

It is clear from the second line that $\kappa_b = A_1$ and $\kappa_g = A_2 - A_1$. It is clear from the third line that $A_1 \pm A_2 \geq 0$ is required for stability. It follows that $2\kappa_b \geq -\kappa_g \geq 0$.

Of course, all this makes sense only if the higher-order terms in Eq. (2.108) are smaller than the ones we have retained. To address this question, we must comment on the scale of the moduli κ_i. The origin of the bending energy is the elastic energy associated with material strains inside the bilayer. These strains scale with D/R, where D is the thickness of the plasma membrane. The scale for the strain-energy density is set by K_A, so we expect indi-

---

44) At order $1/R^3$, there are four additional terms, $(C_1 + C_2)^3$ and $C_1 C_2 (C_1 + C_2)^2$ plus the coefficients of the two even symmetric fourth-order terms, $E_1(x_1^4 + x_2^4)$ and $E_2 x_1^2 x_2^2$.

vidual terms in Eq. (2.108) to go as $K_A(D/R)^n$. Thus, for example, at second order in $1/R$, we expect $\kappa_b \sim \kappa_g \sim D^2 K_A \sim 8 \times 10^{-18}$ J, with higher-order terms smaller by the generic ratio D/R.[45]

B.2
Local Elastic Energy of Stretch and Shear

The argument starts at Eq. (2.26) and asks what parameters describe the local elastic deformation and how does one characterize the dependence of the local elastic energy density f_{ms} on those parameters in such a way as to respect rotation invariance and membrane isotropy. The discussion is a two-dimensional version of the elasticity theory covered in standard texts such as Ogden (1984), Mase and Mase (1999), Başar and Weichert (2000) and Holzapfel (2000). It occurs in treatments of membrane elasticity such as Evans and Skalak (1980).

Consider a particular point $\mathbf{r_0}$ of S_0 which maps under elastic deformation to a point $\mathbf{r}$ of S. Any infinitesimal 2D neighborhood of undeformed membrane skeleton about $\mathbf{r_0}$ maps continuously to a corresponding infinitesimal neighborhood (generally deformed) of $\mathbf{r}$. Let (a_1, a_2) and (x_1, x_2) be local orthogonal coordinates on initial and final patches, centered on $\mathbf{r_0}$ and $\mathbf{r}$, respectively. Any mapping of the undeformed skeleton S_0 onto the surface S is characterized locally by a linear map $\mathbf{x} = \mathbf{Ma}$ (i.e., $x_i = M_{ij}a_j$, with the usual summation convention), which we shall refer to as the deformation matrix. The 2×2 matrix $\mathbf{M}$ varies, generally, from point to point over the skeleton and may be regarded equivalently as a function of $\mathbf{r_0}$ or of $\mathbf{r}$. We suppose that the mapping $S_0 \to S$ is one-to-one. It follows that $\mathbf{M}$ is real and nonsingular. We can assume without loss of generality that the coordinate axes have been chosen so that $\det \mathbf{M} > 0$. Note that $\mathbf{M}$ is not generally symmetric, so that, although it has eigenvectors, those eigenvectors cannot be assumed orthogonal. A sequence of two successive skeletal deformations, $\mathbf{M_1}$ followed by $\mathbf{M_2}$, is equivalent to the single deformation $\mathbf{M_2 M_1}$, which expresses the composition law for local deformations.

Because the mapping is nonsingular, the polar decomposition theorem (Ogden 1984) guarantees that $\mathbf{M}$ can be represented uniquely in either of two alternative forms, $\mathbf{M} = \mathbf{RU} = \mathbf{VR}$, where $\mathbf{R}$ is a pure 2D rotation ($\mathbf{R} = \begin{bmatrix} \cos\theta & \sin\theta \\ -\sin\theta & \cos\theta \end{bmatrix}$, $\mathbf{R}^{-1} = \mathbf{R}^T$) and the matrices $\mathbf{U}$ and $\mathbf{V}$ are real, symmetric and positive definite. Because it is real symmetric, $\mathbf{U}$ can be diagonalized by an orthogonal transformation of the form, $\mathbf{S}^T \mathbf{U} \mathbf{S} = \begin{bmatrix} \lambda_1 & 0 \\ 0 & \lambda_2 \end{bmatrix}$,

---

[45] This estimate of κ_b is a bit high compared to the measured value (see Table 2.5). The reason lies in certain numerical factors to which this crude dimensional argument does not do adequate justice (Wortis and Evans 1997).

where $\mathbf{S}$ is a rotation matrix built from the eigenvectors of $\mathbf{U}$ and the eigenvalues λ_i are positive (and similarly for $\mathbf{V}$). Thus,

$$\mathbf{M} = \mathbf{RS} \begin{bmatrix} \lambda_1 & 0 \\ 0 & \lambda_2 \end{bmatrix} \mathbf{S}^T, \qquad (2.110)$$

showing that $\mathbf{M}$ can always be regarded as the result of a rotation followed by a simple stretch/compression along orthogonal axes followed by a second, independent rotation. The eigenvalues λ_i are just the principle extension ratios introduced above Eq. (2.24). Note that

$$\det \mathbf{M} = \lambda_1 \lambda_2, \qquad (2.111)$$

since the rotation matrices do not contribute.

The assumption, Eq. (2.26), of a local elastic energy density means that f_{ms} depends locally on $\mathbf{M}$ at each point of the membrane skeleton, $f_{\text{ms}}[\mathbf{M}]$. However, mappings of $\mathbf{M}$ which are pure rotations (without stretch/compression) are "isometric" in the sense that they do not change distances and should not cost elastic energy. Thus, we impose the condition that $f_{\text{ms}}[\mathbf{R}] = f_{\text{ms}}[\mathbf{1}] = 0$ and, more generally, that

$$f_{\text{ms}}[\mathbf{RM}] = f_{\text{ms}}[\mathbf{M}], \qquad (2.112)$$

which states that the energy density associated with a general local deformation $\mathbf{M}$ is unchanged by a subsequent arbitrary rotation $\mathbf{R}$.[46] Equation (2.112) is a functional equation which restricts the form of the dependence $f_{\text{ms}}[\mathbf{M}]$.[47] The unique solution of this functional equation is the requirement that the energy density must depend on the combination $\mathbf{M}^T \mathbf{M}$, $f_{\text{ms}}[\mathbf{M}^T \mathbf{M}]$. Note that $\mathbf{M}^T \mathbf{M}$ is symmetric, so that the net result of applying the condition (2.112) is to reduce the number of real variables on which f_{ms} can depend from the four parameters of $\mathbf{M}$ to the three parameters of $\mathbf{M^T M}$, which may be thought of as the two principle extension ratios and the rotation angle associated with S in Eq. (2.110). It is common in the literature to use in the place of $\mathbf{M}^T \mathbf{M}$ the equivalent quantity $u \equiv \frac{1}{2}(\mathbf{M}^T \mathbf{M} - 1)$, which is called the (Cauchy) strain tensor and has the form,

$$u_{ij} = \frac{1}{2}\left(\frac{\partial x_k}{\partial a_i}\frac{\partial x_k}{\partial a_j} - \delta_{ij}\right). \qquad (2.113)$$

Note that the strain tensor measures the change in length of the infinitesimal vector $d\mathbf{a}$ as it is transformed under deformation into the new vector $d\mathbf{x}$,

$$2d\mathbf{a} \cdot u \cdot d\mathbf{a} = 2u_{ij} da_i da_j = dx_i dx_i - da_i da_i = dl^2 - dl_0^2. \qquad (2.114)$$

[46] This condition is sometimes referred to as "objectivity" in the elasticity literature.
[47] A direct analogue is the requirement, $q(\mathbf{Rv}) = q(\mathbf{v})$, that defines a scalar function q of a vector $\mathbf{v}$. The solution of this functional equation is that q must depend on $\mathbf{v}$ only via $\mathbf{v}^2 \equiv \mathbf{v}^T \mathbf{v}$.

The condition of membrane isotropy, which has not yet been applied, takes the form,

$$f_{\rm ms}[\mathbf{MR}] = f_{\rm ms}[\mathbf{M}], \tag{2.115}$$

similar to Eq. (2.112) but now with the rotation preceding the deformation. Combining this with the consequence of rotation invariance gives,[48]

$$f_{\rm ms}[\mathbf{R}^T\mathbf{M}^T\mathbf{M}\mathbf{R}] = f_{\rm ms}[\mathbf{M}^T\mathbf{M}], \tag{2.116}$$

for an arbitrary rotation $\mathbf{R}$, which means that $f_{\rm ms}$ can only depend on rotational invariants of the symmetric matrix $\mathbf{M}^T\mathbf{M}$. These invariants can be taken as

$$\operatorname{tr}\mathbf{M}^T\mathbf{M} = \lambda_1^2 + \lambda_2^2 \quad \text{and} \quad \det\mathbf{M}^T\mathbf{M} = (\det\mathbf{M})^2 = \lambda_1^2\lambda_2^2. \tag{2.117}$$

They are equivalent to the area and shear strains, Eqs. (2.24) and (2.25), in the form,

$$\alpha = \det\mathbf{M} - 1 \quad \text{and} \quad \beta = \frac{1}{2\det\mathbf{M}}\left(\operatorname{tr}\mathbf{M}^T\mathbf{M} - 2\det\mathbf{M}\right), \tag{2.118}$$

which have been chosen for convenience to vanish in the unstrained state $\lambda_1 = \lambda_2 = 1$.

We have established at this point the general form, $f_{\rm ms}(\alpha, \beta)$, for the local elastic energy. Convention assigns zero energy density to the unstrained state, $f_{\rm ms}(0,0) = 0$, and suggests a series development in positive powers of α and β. The further requirement that the unstrained state should be the energy minimum eliminates any term linear in α, so the leading terms in the expansion are β and α^2, both of which are quadratic in the deviations $\delta_i = \lambda_i - 1$ from the unstrained state. These two lowest-order terms are the basis of the "linear" elastic model, Eq. (2.27), which defines the linear moduli μ and K_α. The higher-order elastic terms of Eq. (2.28) simply extend the power series.

Appendix C
Differential Geometry and Coordinate Transformations

This appendix contains a review of some basic results from differential geometry and curvilinear coordinate transformations. It provides notation and results useful for other parts of this article. Good texts on this subject include do Carmo (1976), Kreyszig (1991) and Spivak (1979). In what follows

---

[48] It is easy enough to directly identify $\det\mathbf{M}$ as one invariant satisfying both Eqs. (2.112) and (2.115), since $\det\mathbf{M} = \det\mathbf{RM} = \det\mathbf{MR}$. The problem is the second invariant. Although $\operatorname{tr}\mathbf{RM} = \operatorname{tr}\mathbf{MR}$, it is *not* generally true that $\operatorname{tr}\mathbf{RM} = \operatorname{tr}\mathbf{M}$. It is for this reason that it is necessary for the argument to identify the dependence on $\mathbf{M}^T\mathbf{M}$ before applying Eq. (2.115).

we use bold face for 3D (physical) vectors and adopt the convention that repeated indices are summed unless explicitly indicated to the contrary.

C.1
Basic Results from Differential Geometry

Consider a two-dimensional manifold S embedded in three-dimensional space and labeled by a general set of coordinates (s^1, s^2) (see Fig. 2.51).[49] Suppose that the (3D) vector function $\mathbf{R}(s^1, s^2)$ gives the position of each surface point. Then,

$$\mathbf{Y}_\alpha \equiv \mathbf{Y}_\alpha(s^1, s^2) \equiv \frac{\partial \mathbf{R}}{\partial s^\alpha} \equiv \partial_\alpha \mathbf{R} \qquad (2.119)$$

defines a pair of (generally unnormalized) tangent vectors at (s^1, s^2). If S is closed, as it is for red-cell shapes, then we will assume for simplicity that the labeling of the coordinates has always been chosen locally so that $\mathbf{Y}_1 \times \mathbf{Y}_2$ points along the outwardly directed unit normal $\hat{\mathbf{n}}$. The length dl of the infinitesimal in-plane vector $\mathbf{dl} = \mathbf{Y}_\alpha ds^\alpha$ satisfies $dl^2 = g_{\alpha\beta} ds^\alpha ds^\beta$ with

$$g_{\alpha\beta} \equiv \mathbf{Y}_\alpha \cdot \mathbf{Y}_\beta, \qquad (2.120)$$

which identifies $g_{\alpha\beta} = (\mathbf{g})_{\alpha\beta}$ as the metric tensor and shows that $g_{\alpha\beta}$ is symmetric and $g_{\alpha\alpha}$ (not summed) is positive. It will be convenient to denote the inverse of the metric tensor by raised indices, $g^{\alpha\beta} \equiv (\mathbf{g}^{-1})_{\alpha\beta}$. It is easy to show that $|\mathbf{Y}_1 \times \mathbf{Y}_2|^2 = \det \mathbf{g} \equiv g$, so $g > 0$ and

$$\hat{\mathbf{n}} = \frac{1}{\sqrt{g}} \mathbf{Y}_1 \times \mathbf{Y}_2. \qquad (2.121)$$

The element of area is given by

$$dA = |\mathbf{Y}_1 \times \mathbf{Y}_2| ds^1 ds^2 = \sqrt{g}\, ds^1 ds^2. \qquad (2.122)$$

It will be useful in what follows to introduce the antisymmetric matrices (Lomholt and Miao 2006),

$$\epsilon_{\alpha\beta} = \sqrt{g} \begin{bmatrix} 0 & 1 \\ -1 & 0 \end{bmatrix} \text{ and } \epsilon^{\alpha\beta} = \frac{1}{\sqrt{g}} \begin{bmatrix} 0 & 1 \\ -1 & 0 \end{bmatrix}, \qquad (2.123)$$

which satisfy $\epsilon^{\alpha\beta} \epsilon_{\beta\gamma} = -\delta^\alpha_\gamma$. It is not hard to verify that $\epsilon^{\alpha\beta} = g^{\alpha\sigma} \epsilon_{\sigma\tau} g^{\tau\beta}$, so the metric tensor may be used to raise and lower indices of the ϵ tensors in a fluent manner. This notation allows us to write compactly,

$$\mathbf{Y}_\alpha \times \mathbf{Y}_\beta = \epsilon_{\alpha\beta} \hat{\mathbf{n}}. \qquad (2.124)$$

---

[49] We use superscripted coordinates because the differentials (ds^1, ds^2) will turn out to be contravariant (see below). The full rationale for sub- and superscript notation for co- and contravariant quantities is given following Eq. (2.136). However, we will use the notation consistently throughout this Appendix, even before its significance has been established.

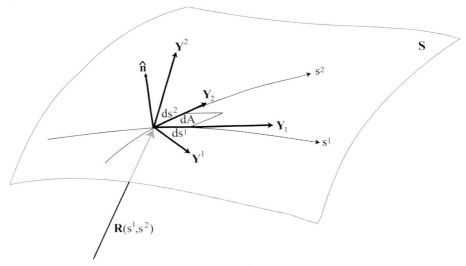

Fig. 2.51 General curvilinear coordinates (s^1, s^2) on the surface S defined by $\mathbf{R}(s^1, s^2)$ embedded in 3D space. The unit normal vector $\hat{\mathbf{n}}$ points outward. The area element dA is defined by the infinitesimal coordinate increments (ds^1, ds^2). The covariant tangent vectors $\mathbf{Y}_1, \mathbf{Y}_2$ are directed locally along the coordinate axes. The corresponding contravariant vectors $\mathbf{Y}^1, \mathbf{Y}^2$ defined by Eq. (2.126) also lie in the tangent plane but are directed perpendicularly to the edges of dA in the sense of increasing s^α.

It is important to introduce the conjugate (contravariant) pair of in-plane vectors $\mathbf{Y}^\alpha \equiv g^{\alpha\beta} \mathbf{Y}_\beta$, which satisfy

$$\mathbf{Y}^\alpha \cdot \mathbf{Y}^\beta = g^{\alpha\beta} \text{ and } \mathbf{Y}_\alpha \cdot \mathbf{Y}^\beta = \delta_\alpha^\beta, \tag{2.125}$$

where (as in the text) we adopt the convention of writing the arguments of the δ-function up or down in such a way as to preserve covariant fluency. Any vector in the tangent plane can be expressed as a linear combination of $\mathbf{Y}_1$ and $\mathbf{Y}_2$ or, alternatively, of $\mathbf{Y}^1$ and $\mathbf{Y}^2$. Finally, it will be useful to have

$$\left(\hat{\mathbf{n}} \times \mathbf{Y}_\alpha\right) = \epsilon_{\alpha\beta} \mathbf{Y}^\beta \text{ and } \left(\hat{\mathbf{n}} \times \mathbf{Y}^\alpha\right) = \epsilon^{\alpha\beta} \mathbf{Y}_\beta. \tag{2.126}$$

This completes the discussion of geometry at a single point. We now pass to the discussion of the way these quantities change in moving to nearby points.

Coordinate derivatives of the tangent vectors have, in general, both in-plane and out-of-plane components, so it is always possible to write,

$$\mathbf{Y}_{\alpha\beta} \equiv \partial_\beta \mathbf{Y}_\alpha = \partial_\alpha \mathbf{Y}_\beta = \partial_\alpha \partial_\beta \mathbf{R} = \Gamma^\gamma_{\alpha\beta} \mathbf{Y}_\gamma - C_{\alpha\beta} \hat{\mathbf{n}}, \tag{2.127}$$

in which the coefficients $\Gamma^\gamma_{\alpha\beta}$ and $C_{\alpha\beta}$ are manifestly symmetric under the interchange $\alpha \leftrightarrow \beta$. The quantities $\Gamma^\gamma_{\alpha\beta}$ are called Christoffel symbols. They are directly related to coordinate derivatives of the metric tensor, since

$$\partial_\gamma g_{\alpha\beta} = \partial_\gamma \left(\mathbf{Y}_\alpha \cdot \mathbf{Y}_\beta \right) = \mathbf{Y}_{\alpha\gamma} \cdot \mathbf{Y}_\beta + \mathbf{Y}_\alpha \cdot \mathbf{Y}_{\beta\gamma} = \Gamma^\delta_{\alpha\gamma} g_{\delta\beta} + \Gamma^\delta_{\beta\gamma} g_{\delta\alpha}, \tag{2.128}$$

which leads to the useful relation,

$$\partial_\alpha g = 2g \Gamma^\beta_{\alpha\beta}. \tag{2.129}$$

Equation (2.128) can be inverted to give

$$\Gamma^\gamma_{\alpha\beta} = \frac{1}{2} \left(\partial_\alpha g_{\beta\delta} + \partial_\beta g_{\alpha\delta} - \partial_\delta g_{\alpha\beta} \right) g^{\delta\gamma}. \tag{2.130}$$

We are now in a position to calculate the coordinate derivatives of the normal vector, which are related to the curvature tensor. By taking directly the derivative of Eq. (2.121) and using Eq. (2.129), we find $\hat{\mathbf{n}} \cdot \partial_\alpha \hat{\mathbf{n}} = 0$, so

$$\partial_\alpha \hat{\mathbf{n}} = -\frac{1}{\sqrt{g}} \left(C_{1\alpha} \hat{\mathbf{n}} \times \mathbf{Y}_2 + C_{2\alpha} \mathbf{Y}_1 \times \hat{\mathbf{n}} \right) = C_{\alpha\beta} \mathbf{Y}^\beta = C_\alpha^{\ \beta} \mathbf{Y}_\beta, \tag{2.131}$$

where $C_\alpha^{\ \beta} \equiv C_{\alpha\gamma} g^{\gamma\beta} \equiv (\mathbf{C})_\alpha^{\ \beta}$. Equation (2.131) is called the Weingarten relation and defines the usual geometric curvature tensor, $\mathbf{C}$, whose eigenvalues can be identified (see below) with the principal curvatures introduced in Section 2.3.4. Note that, although $C_{\alpha\beta}$ is always symmetric under $\alpha \leftrightarrow \beta$, the curvature tensor is not. On the other hand, symmetry does guarantee that $C_\alpha^{\ \beta} = C^\beta_{\ \alpha}$, so we can write C_α^β, with the indices aligned, without ambiguity. It is convenient to define

$$K \equiv \det \mathbf{C} = C_1 C_2 \qquad \text{(Gaussian curvature)} \tag{2.132}$$

and

$$H \equiv \frac{1}{2} \operatorname{tr} \mathbf{C} = \frac{1}{2}(C_1 + C_2) \qquad \text{(mean curvature)}. \tag{2.133}$$

Because $\mathbf{C}$ is a 2×2 matrix, it follows that

$$\mathbf{C}^2 - 2H\mathbf{C} + K\mathbf{1} = 0, \text{ so } \operatorname{tr} C^2 = 4H^2 - 2K. \tag{2.134}$$

Finally, we will need an additional relation connected to the third derivatives, $\partial_\alpha \partial_\beta \partial_\gamma \mathbf{R}$. Because these derivatives are manifestly invariant under permutation of indices, we find by using Eq. (2.127)

$$\partial_\alpha \left(\Gamma^\delta_{\beta\gamma} \mathbf{Y}_\delta - C_{\beta\gamma} \hat{\mathbf{n}} \right) = \partial_\beta \left(\Gamma^\delta_{\alpha\gamma} \mathbf{Y}_\delta - C_{\alpha\gamma} \hat{\mathbf{n}} \right) = \partial_\gamma \left(\Gamma^\delta_{\alpha\beta} \mathbf{Y}_\delta - C_{\alpha\beta} \hat{\mathbf{n}} \right), \tag{2.135}$$

which is one form of the so-called Codazzi relations.

C.2
Coordinate Transformations and Covariant Notation

It will be useful now to discuss the manner in which various quantities transform under (locally) nonsingular coordinate changes, $s'^\alpha = s'^\alpha(s^1, s^2)$ (we assume for simplicity that the direction of the local normal $\hat{\mathbf{n}}$ is not changed). Under coordinate changes some quantities remain invariant in the sense that $Q'(s'^1, s'^2) = Q(s^1, s^2)$. Such quantities are said to be scalar under coordinate transformation. This designation is unrelated to spatial-rotation properties: normal scalars, vectors, and so on, can all be "scalar" under coordinate change. Note that physical fields (densities, orientational fields, etc.) are scalar in this sense, since they do not change value when the labeling of the surface is reparametrized. By contrast, sub- or superscripted quantities like $\mathbf{Y}_\alpha$ and $C_{\alpha\beta}$ vary under coordinate change. Such indexed quantities are defined to be "tensors" when their transformation under coordinate change is controlled by the Jacobian matrix,

$$J_{\alpha\beta} \equiv \frac{\partial s'^\alpha}{\partial s^\beta}, \quad \text{or its inverse,} \quad J^{-1}_{\alpha\beta} \equiv \frac{\partial s^\alpha}{\partial s'^\beta}. \tag{2.136}$$

We adopt the usual convention of using lower and upper indices to denote, respectively, co- and contravariant "tensor" quantities in the sense that, under coordinate transformation, first-rank (rank 1, single-index) tensor quantities transform as

$$Q'_\alpha = Q_\beta J^{-1}_{\beta\alpha} \quad \text{and} \quad Q'^\alpha = J_{\alpha\beta} Q^\beta, \tag{2.137}$$

and similarly with additional indices (higher-rank tensors). It is easy to show that the tangent vectors $\mathbf{Y}_\alpha$ and the metric tensor $g_{\alpha\beta}$ are covariant, while the infinitesimal displacement vector $ds^\alpha = (ds^1, ds^2)$ and the inverse of the metric tensor $g^{\alpha\beta}$ are contravariant. Furthermore, we verify that the metric tensor and its inverse may be used to raise and lower indices consistently, according to

$$Q_\alpha = g_{\alpha\beta} Q^\beta \quad \text{and} \quad Q^\alpha = g^{\alpha\beta} Q_\beta, \tag{2.138}$$

so that, for example, $\mathbf{Y}^\alpha = g^{\alpha\gamma} \mathbf{Y}_\gamma$ with $\mathbf{Y}^\alpha \cdot \mathbf{Y}_\beta = g^{\alpha\gamma} \mathbf{Y}_\gamma \cdot \mathbf{Y}_\beta = g^{\alpha\gamma} g_{\gamma\beta} = \delta^\alpha_\beta$. Note that mixed tensors of second rank have the property that, under coordinate change, they transform according to

$$M'^\alpha{}_\beta = J_{\alpha\bar{\alpha}} M^{\bar{\alpha}}{}_{\bar{\beta}} J^{-1}_{\bar{\beta}\beta} \quad \text{and} \quad M'_\alpha{}^\beta = (J^T)^{-1}_{\alpha\bar{\alpha}} M_{\bar{\alpha}}{}^{\bar{\beta}} J^T_{\bar{\beta}\beta}, \tag{2.139}$$

where T denotes the transpose. In either of these forms, it follows from Eq. (2.139) that the trace, the determinant and all the eigenvalues of M are invariant under coordinate changes, a property we will need below.

The transformation behavior of derivatives requires particular care. Differentiation of a scalar is straightforward,

$$\partial'_\alpha Q' = \frac{\partial}{\partial s'^\alpha} Q = \frac{\partial s^\beta}{\partial s'^\alpha} \frac{\partial}{\partial s^\beta} Q = (\partial_\beta Q) J^{-1}_{\beta\alpha}, \tag{2.140}$$

so that $\partial_\alpha Q$ is a covariant rank 1 tensor according to the definition Eq. (2.137). Eq. (2.119) is an example (note that $\mathbf{R}$ is a scalar). On the other hand, differentiation of a rank 1 tensor does not produce a tensor quantity, since, for example,

$$\partial'_\alpha Q'_\beta = \partial'_\alpha \left(Q_{\bar\beta} J^{-1}_{\bar\beta\beta} \right) = \left(\partial'_\alpha Q_{\bar\beta} \right) J^{-1}_{\bar\beta\beta} + Q_{\bar\beta} \partial'_\alpha J^{-1}_{\bar\beta\beta}$$

$$= \left(\partial_{\bar\alpha} Q_{\bar\beta} \right) J^{-1}_{\bar\alpha\alpha} J^{-1}_{\bar\beta\beta} + Q_{\bar\beta} \frac{\partial^2 s^{\bar\beta}}{\partial s'^\alpha \partial s'^\beta}. \qquad (2.141)$$

The first term on the right looks like a second-rank covariant tensor but the second term is non-tensorial, so we must conclude that $\partial_\alpha Q_\beta$ is not a tensor quantity. It turns out, however, that the modified expression,

$$D_\alpha Q_\beta \equiv \partial_\alpha Q_\beta - \Gamma^\gamma_{\alpha\beta} Q_\gamma, \qquad (2.142)$$

is a second-rank covariant tensor. Proving this requires verifying that

$$D'_\alpha Q'_\beta \equiv \partial'_\alpha Q'_\beta - \Gamma'^\gamma_{\alpha\beta} Q'_\gamma$$

$$= \left(\partial_{\bar\alpha} Q_{\bar\beta} - \Gamma^{\bar\gamma}_{\bar\alpha\bar\beta} Q_{\bar\gamma} \right) J^{-1}_{\bar\alpha\alpha} J^{-1}_{\bar\beta\beta} \equiv D_{\bar\alpha} Q_{\bar\beta} J^{-1}_{\bar\alpha\alpha} J^{-1}_{\bar\beta\beta}. \qquad (2.143)$$

Using $Q'_\beta = Q_{\bar\beta} J^{-1}_{\bar\beta\beta}$ from (2.137) and $\partial'_\alpha = J^{-1}_{\bar\alpha\alpha} \partial_{\bar\alpha}$ converts (2.143) into the Q-independent condition,

$$\Gamma'^\gamma_{\alpha\beta} = J_{\gamma\bar\gamma} \Gamma^{\bar\gamma}_{\bar\alpha\bar\beta} J^{-1}_{\bar\alpha\alpha} J^{-1}_{\bar\beta\beta} + J_{\gamma\bar\gamma} \frac{\partial s^{\bar\gamma}}{\partial s'^\alpha \partial s'^\beta}, \qquad (2.144)$$

which in turn follows from Eq. (2.130). Note that Eq. (2.144), which gives the transformation properties of the Christoffel symbols under coordinate change, shows that $\Gamma^\gamma_{\alpha\beta}$ fails to transform as a (mixed) third-rank tensor because of a second-derivative term similar to that which appears in (2.141). Indeed, it is a cancellation between these two non-tensorial terms which conspires to allow $D_\alpha Q_\beta$ to transform as a (covariant) tensor. In a similar manner, one shows that

$$D_\alpha Q^\beta \equiv \partial_\alpha Q^\beta + \Gamma^\beta_{\alpha\gamma} Q^\gamma \qquad (2.145)$$

transforms as a mixed tensor, covariant in the first index but contravariant in the second. A useful special case is the identity,

$$D_\alpha Q^\alpha = \frac{1}{\sqrt{g}} \partial_\alpha \left(\sqrt{g}\, Q^\alpha \right), \qquad (2.146)$$

which make use of Eq. (2.129). Equations (2.142) and (2.145) define the so-called covariant derivatives of co- and contravariant first-rank tensors, respectively. Covariant derivatives of higher-rank tensors may be handled in a similar manner. Thus,

$$D_\alpha Q_{\beta\gamma} \equiv \partial_\alpha Q_{\beta\gamma} - \Gamma^\delta_{\alpha\beta} Q_{\delta\gamma} - \Gamma^\delta_{\alpha\gamma} Q_{\beta\delta} \qquad (2.147)$$

and

$$D_\alpha Q^{\beta\gamma} \equiv \partial_\alpha Q^{\beta\gamma} + \Gamma^\beta_{\alpha\delta} Q^{\delta\gamma} + \Gamma^\gamma_{\alpha\delta} Q^{\beta\delta} \qquad (2.148)$$

can be shown to transform as third-rank tensors. The general rule for all mixed and higher-order tensors is that each lower (covariant) index comes in with a $-\Gamma$ term, while each upper (contravariant) index comes in with a $+\Gamma$. Combining all these definitions (including $D_\alpha Q \equiv \partial_\alpha Q$ for scalars) produces a general definition of a covariant derivative operation D_α which acts on tensor quantities of rank n to consistently produce corresponding tensor quantities of rank $(n+1)$ with an additional covariant index α. The operation of the covariant derivative thus defined obeys the usual chain rule, so that, for example, $D_\alpha(QP_\beta) = (D_\alpha Q)P_\beta + Q(D_\alpha P_\beta)$ and $D_\alpha(Q_{\beta\gamma}P^{\gamma\delta}) = (D_\alpha Q_{\beta\gamma})P^{\gamma\delta} + Q_{\beta\gamma}(D_\alpha P^{\gamma\delta})$, and so on, as long as the contracted indices are an upper-lower pair.

Various formulae may be conveniently written using the covariant derivative. Thus,

$$D_\alpha \mathbf{Y}_\beta = -C_{\alpha\beta}\hat{\mathbf{n}} \quad \text{(Eq. (2.127))} \quad (2.149)$$

$$D_\alpha \hat{\mathbf{n}} = C_\alpha^{\ \beta}\mathbf{Y}_\beta \quad \text{(Weingarten)} \quad (2.150)$$

$$D_\alpha g_{\beta\gamma} = D_\alpha g^{\beta\gamma} = 0 \quad \text{(Eq. (2.128))} \quad (2.151)$$

$$D_\alpha C_{\beta\gamma} = D_\beta C_{\alpha\gamma} \quad \text{(Codazzi)} \quad (2.152)$$

$$D_\alpha \epsilon_{\beta\gamma}\, D_\alpha \epsilon_{\beta\gamma} = 0 \quad (2.153)$$

It is useful to define $D^\alpha \equiv g^{\alpha\beta}D_\beta$ and

$$\Delta \equiv D_\alpha D^\alpha = \frac{1}{\sqrt{g}}\partial_\alpha(\sqrt{g}\,\partial^\alpha), \quad (2.154)$$

which is called the Laplace–Beltrami operator.

C.3
Physical Quantities

Physical quantities are invariant under coordinate change. We will be dealing with physical fields which live on the 2D surface S defined by $\mathbf{R} = \mathbf{R}(s^1, s^2)$ embedded in 3D space. Suppose that $Q(s^1, s^2)$ is a physical scalar field over S. It is natural to define the generalized in-plane gradient of Q by the property,

$$dQ \equiv \partial_\alpha Q\, ds^\alpha \equiv \nabla Q \cdot d\mathbf{R}, \quad (2.155)$$

for arbitrary infinitesimal surface displacement $d\mathbf{R} = \mathbf{Y}_\alpha ds^\alpha$. It follows that

$$\nabla Q = (\partial_\alpha Q)\mathbf{Y}^\alpha = (D_\alpha Q)\mathbf{Y}^\alpha = (D^\alpha Q)\mathbf{Y}_\alpha = (\mathbf{Y}^\alpha D_\alpha)Q, \quad (2.156)$$

which is a physical vector in the tangent plane and explicitly invariant under coordinate transformations. The last equality of Eq. (2.156) suggests defining the physical-vector differential operator,

$$\nabla = \mathbf{Y}^\alpha D_\alpha = \mathbf{Y}_\alpha D^\alpha, \tag{2.157}$$

Now, suppose in a similar way that $\mathbf{W}(s^1, s^2)$ is a physical vector field over S. Generally $\mathbf{W}$ may have components in the tangent plane and normal to it,

$$\mathbf{W} = W^\beta \mathbf{Y}_\beta + W_\perp \hat{\mathbf{n}}. \tag{2.158}$$

Coordinate derivatives of $\mathbf{W}$ are easy to calculate using the chain rule and Eqs. (2.149)–(2.152) above,

$$\partial_\alpha \mathbf{W} = D_\alpha \mathbf{W} = (D_\alpha W^\beta)\mathbf{Y}_\beta + W_\perp C_\alpha^{\ \beta}\mathbf{Y}_\beta + \big(-C_{\alpha\beta}W^\beta + \partial_\alpha W_\perp\big)\hat{\mathbf{n}}. \tag{2.159}$$

Combining Eqs. (2.157) and (2.159), we evaluate the generalized divergence of $\mathbf{W}$,

$$\nabla \cdot \mathbf{W} = (\mathbf{Y}^\alpha D_\alpha)\mathbf{W} = D_\alpha W^\alpha + 2HW_\perp. \tag{2.160}$$

An important example of this relation is when $\mathbf{W} = \nabla Q$, so $(\nabla Q)^\alpha = D^\alpha Q$ and $\nabla \cdot \nabla Q = \Delta Q$.

An important relation which we shall quote but not prove (Spivak 1965) is a generalization of Gauss's law. Suppose that Σ is a patch of curved surface with perimeter $\partial\Sigma$. Let dl be an element of $\partial\Sigma$ and $\hat{\mathbf{p}} \equiv p_\alpha \mathbf{Y}^\alpha$ be the in-plane unit vector normal to dl. Suppose that $\mathbf{T}(\hat{\mathbf{p}})$ is a 3D physical vector with the invariant form $\mathbf{T}(\hat{\mathbf{p}}) = \mathbf{T}^\alpha p_\alpha$, then

$$\int_{\partial\Sigma} dl\, \mathbf{T}(\hat{\mathbf{p}}) = \int_{\partial\Sigma} dl\, \mathbf{T}^\alpha p_\alpha = \int_\Sigma dA\, D_\alpha \mathbf{T}^\alpha. \tag{2.161}$$

Note in this relation that the physical vectors $\mathbf{T}^\alpha$ can have both in-plane and out-of-plane components.

We end this section with a small piece of unfinished business: the geometric identification of the eigenvalues C_γ of the curvature matrix. It follows from Eq. (2.149) that $C_\alpha^{\ \beta}$ is a mixed rank 2 tensor under coordinate transformations, so its eigenvalues are coordinate independent by virtue of Eq. (2.139). Thus, without changing the eigenvalues, we are free to go to coordinates which are locally Cartesian, $g_{\alpha\beta} = \delta_{\alpha\beta}$, so that $C_\alpha^{\ \beta}$ becomes symmetric and then to rotate coordinates to make it diagonal, $C_{\alpha\beta} = C_\alpha \delta_{\alpha\beta}$. In these coordinates Eq. (2.131) now reads $\partial_\alpha \hat{\mathbf{n}} = C_\alpha \mathbf{Y}_\alpha$ (not summed). It then follows that the magnitude of the change $d\mathbf{n}$ in $\hat{\mathbf{n}}$ as s^α increases by ds^α is given by the angular deviation $d\theta$ of the normal, so $dn = C_\alpha ds^\alpha = d\theta = \frac{ds^\alpha}{R_\alpha}$, where R_α is the local principal radius of curvature of S in the direction α (see Eq. (2.17)) and $C_\alpha = \frac{1}{R_\alpha}$.

C.4
Variational Approach to Membrane Mechanics

Although we have focussed on a Newtonian, force-based approach to equilibrium membrane mechanics, it may be useful here to collect some results which find application in the variational approach to vesicle shapes. Suppose that we start with some initial membrane shape $\mathbf{R}(s^1, s^2)$ and consider a new shape derived from it by small variations of the surface in the normal direction,

$$\mathbf{R}' \equiv \mathbf{R} + \delta\mathbf{R} \equiv \mathbf{R} + \eta\hat{\mathbf{n}}, \tag{2.162}$$

where $\eta = \eta(s^1, s^2)$ is small and will be treated to linear order in what follows. The infinitesimally deformed surface $\mathbf{R}'$ is labeled by the same coordinates (s^1, s^2) but has its own geometry, which differs at order η from that of $\mathbf{R}$. Thus, the new tangent vectors are

$$\mathbf{Y}'_\alpha \equiv \mathbf{Y}_\alpha + \delta\mathbf{Y}_\alpha \equiv \partial_\alpha \mathbf{R}' = \partial_\alpha (\mathbf{R} + \delta\mathbf{R})$$
$$= \mathbf{Y}_\alpha + \eta C_\alpha^{\ \beta} \mathbf{Y}_\beta + (\partial_\alpha \eta)\hat{\mathbf{n}}, \tag{2.163}$$

where we have used Eq. (2.131). A short calculation based on Eq. (2.120) then shows that the change in the metric tensor defined by $g'_{\alpha\beta} \equiv g_{\alpha\beta} + \delta g_{\alpha\beta}$ is

$$\delta g_{\alpha\beta} = 2\eta C_{\alpha\beta}, \tag{2.164}$$

from which the change δg in $g \equiv \det \mathbf{g}$ is

$$\delta g = 4\eta H\, g, \tag{2.165}$$

where we have used the general relation,

$$\delta \ln \det M = \operatorname{tr} M^{-1} \delta M. \tag{2.166}$$

It is now easy to calculate the variation δA of the surface area defined by

$$A' \equiv \int_{S'} ds^1 ds^2 \sqrt{g'} \equiv A + \delta A,$$

$$\delta A = \int_S dA\, \eta(2H), \tag{2.167}$$

which gives Eq. (2.21) for the leaflet area difference $\Delta A[S]$ in the special case where $\eta(s^1, s^2)$ is equal to the constant offset D_0 between the neutral planes.

Linear variations of the geometric variables determine the change in the variational functional F_{var}, Eq. (2.38), the stationarity of which is required for mechanical equilibrium. Assume, for example, that F_{pm} has the form of the sum of Eqs. (2.14) and (2.15),

$$F_{\text{pm}} = \int_S dA\, f(H, K) = \int_S dA \left[\frac{\kappa_b}{2}(2H - C_0)^2 + \kappa_g K \right]. \tag{2.168}$$

Then,

$$\delta F_{\text{var}} = \delta F_{\text{pm}} + \Sigma \delta A - \Delta P \delta V$$

$$= \int_S ds^1 ds^2 \sqrt{g}\, [\eta(2Hf) + \delta f] + \Sigma \delta A - \Delta P \delta V, \tag{2.169}$$

with $\delta f = \kappa_b(2H - C_0)\delta(2H) + \kappa_g \delta K$. Thus, to evaluate the variation δF_{var}, all that is needed are the four geometrical variations $\delta A, \delta V, \delta(2H)$ and δK. The first of these is given by Eq. (2.167). The remaining variations may be evaluated by using the techniques outlined above. The results are

$$\delta V = \int_S dA\, \eta \tag{2.170}$$

$$\delta(2H) = -2\eta(2H^2 - K) + g^{\alpha\beta}\Gamma^{\gamma}_{\alpha\beta}\partial_\gamma \eta - g^{\alpha\beta}(\partial_\alpha \partial_\beta \eta) \tag{2.171}$$

$$\delta K = -2\eta H (8H^2 - 3K)$$
$$+ (2H g^{\alpha\beta} + C^{\alpha\beta})(\partial_\alpha \partial_\beta \eta - \Gamma^{\gamma}_{\alpha\beta}\partial_\gamma \eta). \tag{2.172}$$

By use of these expressions in Eq. (2.169), the equilibrium condition may be written in the form,

$$0 = \delta F_{\mathrm{var}}$$
$$= \int_S dA \left[2H \left(\Sigma + \frac{\kappa_b}{2}(2H - C_0)^2 \right) - \Delta \mathcal{M} \right.$$
$$\left. - \mathcal{M}(4H^2 - 2\det \mathbf{C}) - \Delta P \right] \eta + \text{boundary terms}, \tag{2.173}$$

where integration by parts has been used to transfer the derivatives away from the shape variations η. For a closed vesicle, the boundary terms vanish and the shape equation (2.67) reemerges. This variational approach was used in the original derivation by Ou-Yang and Helfrich (1987a) (see also Ou-Yang and Helfrich (1987b), Ou-Yang and Helfrich (1989), Capovilla and Guven (2002) and Fournier (2007)).

Appendix D
Mechanical Equations of Membrane Equilibrium

In this appendix we present details of some of the membrane-mechanics calculations that have been outlined in Sections 2.4.2 and 2.4.3. We will make free use of general curvilinear coordinates and the results of Appendix C. We start by demonstrating the decomposition of boundary stresses and their relation to the stress tensor. In Appendix D.2 we develop the form of the stress tensor for a simplified model of the fluid membrane with bending rigidity and we show how force and moment balance leads to equations identical to those derived variationally by Ou-Yang and Helfrich (1987a). Inclusion of the Gaussian-curvature term is discussed in Appendix D.3. Appendix D.4 develops the treatment of the deformation matrix $\mathbf{M}$ in curvilinear coordinates. These results are applied to the membrane skeleton in Appendix D.5, where they allow the stress tensor to be derived from the elastic-energy functional.

Finally, in Appendix D.6, we derive explicit equilibrium-shape equations for axisymmetric geometry, first for the plasma membrane only and then for the composite membrane.

D.1
Decomposition of the Stress Tensor for Membranes

As in three-dimensional materials, internal stresses in a membrane produce boundary forces which act on an arbitrary membrane patch Σ (Fig. 2.12). As in Section 2.4.2, we define $\mathbf{T}(\hat{\mathbf{p}})$ to be the force per unit length acting locally at a point (s^1, s^2) on the boundary of Σ. The vector $\hat{\mathbf{p}}$ is the outwardly directed in-plane normal to the boundary element locally at (s^1, s^2). The total force on Σ due to the boundary stresses is given by

$$\mathbf{F}_\Sigma = \int_{\partial\Sigma} dl\, \mathbf{T}(\hat{\mathbf{p}}), \tag{2.174}$$

where dl is the arc length along the boundary $\partial\Sigma$. We will now demonstrate that $\mathbf{T}(\hat{\mathbf{p}})$ can be decomposed as in Eq. (2.50),

$$\mathbf{T}(\hat{\mathbf{p}}) = p_\alpha \mathbf{T}^\alpha, \tag{2.175}$$

where $\hat{\mathbf{p}} = p^\alpha \mathbf{Y}_\alpha = p_\alpha \mathbf{Y}^\alpha$ ($\mathbf{Y}_\alpha$ are the local tangent vectors of Appendix C) and the vectors $\mathbf{T}^\alpha$ are independent of $\hat{\mathbf{p}}$.

Consider (Fig. 2.52(a)) a small triangular patch bounded by a curve c and the two coordinate curves, s^α = constant, which cross one another at the point P. We will be taking the limit as c approaches P and the patch becomes small. In this limit, the curvature of the sides contributes only higher-order terms in the (infinitesimal) lengths, dl, dl_1, dl_2, and we may write,

$$\hat{\mathbf{p}}\, dl + \hat{\boldsymbol{\nu}}_\alpha dl_\alpha = \mathbf{0}, \tag{2.176}$$

where $\hat{\mathbf{p}}$ and $\hat{\boldsymbol{\nu}}_\alpha$ are the outward-pointing in-plane unit normals. The normal vectors $\hat{\boldsymbol{\nu}}_\alpha$ are just normalized versions of the contravariant tangent vectors (Fig. 2.51), $\hat{\boldsymbol{\nu}}_1 = -\mathbf{Y}^2/\sqrt{g^{22}}$ and $\hat{\boldsymbol{\nu}}_2 = -\mathbf{Y}^1/\sqrt{g^{11}}$, so, by comparing coefficients of $\mathbf{Y}^2$ in Eq. (2.176), we find $dl_1 = p_2\sqrt{g^{22}}\,dl$ and $dl_2 = p_1\sqrt{g^{11}}\,dl$. In the same limit, force equilibrium of the triangular patch gives us

$$\mathbf{T}(\hat{\mathbf{p}})\,dl + \mathbf{T}(\hat{\boldsymbol{\nu}}_\alpha)\,dl_\alpha = \mathbf{0}. \tag{2.177}$$

Substituting for dl_α, we find

$$\left(\mathbf{T}(\hat{\mathbf{p}}) + \mathbf{T}(\hat{\boldsymbol{\nu}}_1)p_2\sqrt{g^{22}} + \mathbf{T}(\hat{\boldsymbol{\nu}}_2)p_1\sqrt{g^{11}}\right)dl = \mathbf{0}. \tag{2.178}$$

By defining the $\hat{\mathbf{p}}$-independent quantities, $\mathbf{T}^1 = -\mathbf{T}(\hat{\boldsymbol{\nu}}_2)\sqrt{g^{11}}$ and $\mathbf{T}^2 = -\mathbf{T}(\hat{\boldsymbol{\nu}}_1)\sqrt{g^{11}}$, we arrive finally at Eq. (2.175). A further expansion of the tangential components of $\mathbf{T}^\alpha$ in terms of the tangent vectors gives (Eq. (2.51)),

$$\mathbf{T}^\alpha = T^{\alpha\beta}\mathbf{Y}_\beta + T^\alpha_\perp \hat{\mathbf{n}}. \tag{2.179}$$

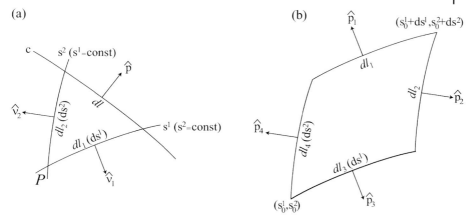

Fig. 2.52 Notation for membrane-patch calculations. (a) Infinitesimal triangular patch bounded on sides 1 and 2 by the coordinate axes and on the third side by the curve c. The corresponding outwardly directed in-plane unit normals $\hat{\nu}_1, \hat{\nu}_2$ and $\hat{\mathbf{p}}$ are shown. This geometry is required for the argument in Sec. D.1. (b) Infinitesimal parallelogram patch bounded on all four sides by the coordinate axes. The outwardly directed in-plane unit normal vectors $\hat{\mathbf{p}}_i$ are shown. This geometry is required for arguments in Sections D.2 and D.5.

D.2
Stress Tensor for the Helfrich Model

In this section we derive the form, Eq. (2.64), of the stress tensor for a uniform isotropic fluid membrane with an isotropic (Helfrich) bending rigidity. There is a large class of microscopic membrane models which lead to the same form of stress tensor, only with different prescriptions for calculating the material parameters κ_b and C_0 in terms of microscopic quantities. All such models have two features. One is membrane fluidity, that is, the absence of resistance to static in-plane shear stresses; the other is some form of stress-density profile across the thickness of the membrane, which is the microscopic origin of the bending moment $\mathcal{M}$, Eq. (2.61). The derivation of Eq. (2.64) is, of course, model dependent. The simplest such model and the one we adopt for our derivation is that of a stack of infinitesimally thin independent sheets or layers, each of width dz and with an isotropic in-plane tension $t(z, s^1, s^2)dz$. We label each layer by its (fixed) perpendicular displacement z from the reference surface $\mathbf{R}(s^1, s^2)$ which will characterize the membrane at the macroscopic level, so that the layer labeled by z has the shape,

$$\mathbf{R}(z, s^1, s^2) = \mathbf{R}(s^1, s^2) + z\hat{\mathbf{n}}(s^1, s^2), \qquad (2.180)$$

parametrized by the same transverse labels (s^1, s^2) in each layer. The overall thickness of the stack (i.e., the range of z) must be small on the scale of all relevant radii of curvature in order that the eventual 2D continuum description (e.g., Eq. (2.14)) should make sense. Note that the position $z = 0$ of the reference surface within the stack is at this point arbitrary. We will find below that there is a natural choice – the so-called "neutral surface" – for this position.

The tension distribution $t(z, s^1, s^2)$ depends on the molecular structure of the membrane and also on the local state of membrane bending. In principle, $t(z, s^1, s^2)$ must emerge from a microscopic calculation based on specific molecular and/or material properties and the way in which these vary both across the thickness (z) of the membrane and along its surface (s^1, s^2). Luckily, however, we will need for our purposes only generic behavior. Note that, in order that the stack be able to produce a local bending moment when deformed, it is necessary that the tension in each layer be able to respond locally to the stretching or compression caused by changes of local curvature, so in general $\partial t/\partial s^\alpha \neq 0$, unlike the situation for Eqs. (2.46) or (2.56). At first sight it may appear surprising that the layer tensions, although isotropic, are generally non-uniform, despite the fact that the model purports to describe a fluid membrane. The point is that the membrane has a thickness on the molecular scale, and its molecular constituents can and do transfer stresses over distances comparable to their dimensions. The z-dependent stresses induced by membrane bending cannot be relieved by fluid motion in individual layers, since the molecules of the membrane span a significant fraction of its thickness. Thus, the non-uniformity in $t(z, s^1, s^2)$ results from lateral non-uniformity of the bending deformations of the membrane and the distribution in z of the bending stresses. For similar reasons, there are also forces acting between the layers in the normal direction, the net effect of which is to produce a non-zero normal component of the stress tensor $\mathbf{T}^\alpha_\perp$, which was absent for the soap film, Eq. (2.54).

The variation $t(z)$ across the thickness of the stack must have two generic properties in order that the membrane mechanics should correspond to those of the Helfrich model. First, there must be a state of relaxed equilibrium (neutral) whenever the mean curvature takes on a value such that $2H = C_0$. This reflects the existence of a spontaneous curvature. Note that even in this "relaxed" state there is generally a non-trivial stress profile $t_0(z)$ across the membrane stack, for example, if the lipid heads are large and the tails small, then the pressure will be large in the head region (i.e., less tension) and small in the tail region (more tension). Relaxed equilibrium means that there is no local bending moment in this state, so $\int dz\, z t_0(z) = 0$, a condition which requires a specific choice for $z = 0$ called the neutral surface. Second, when $2H$ moves away from this relaxed value, the stresses in the membrane must scale as $2H - C_0$, so that overall

$$t(z) = t_0(z) + (2H - C_0)t_1(z) \qquad (2.181)$$

and

$$\mathcal{M} \equiv \int dz\, zt(z) = \kappa_{\rm b}(2H - C_0) \quad \text{with} \quad \kappa_{\rm b} \equiv \int dz\, zt_1(z), \quad (2.182)$$

which is the same as Eq. (2.61) of the text, only now with an explicit prescription giving the bending modulus $\kappa_{\rm b}$ in terms of the stress profile. It is not obvious that the change in the membrane stress profile should scale only or even principally with the isotropic quantity $(2H - C_0)$. Indeed, other measures of membrane geometry may also appear (one of which we will consider in Appendix D.3). However, if terms beyond Eq. (2.181) do occur, then the membrane cannot be described in terms of the spontaneous curvature model, Eq. (2.14).

We turn now to the geometry of the layered stack. Note that, if D is the membrane thickness, then the stress changes in the membrane are expected to be proportional to $D(2H - C_0)$, a ratio which must be small in order that a two-dimensional representation of the membrane should make sense. On this basis we expect $t_1 \sim t_0 D$. A simple model would be $t(z) = t_0(z)[1 + A(2H - C_0)z]$; however, this is not required for what follows. Of course, Equation (2.182) contains only the first term in what is more generally an expansion in powers of the small ratio $D(2H - C_0)$. The higher terms are expected to be small and we will neglect them in the following, thus justifying an expansion in powers of $z\mathbf{C}$.

Because of the offset from $\mathbf{R}(s^1, s^2)$, the differential geometry of layer z will differ from that of the reference surface. Thus, although the local normals agree, $\hat{\mathbf{n}}(z) = \hat{\mathbf{n}}$, the tangent vectors differ,

$$\mathbf{Y}_\alpha(z) = \frac{\partial \mathbf{R}(z)}{\partial s^\alpha} = \mathbf{Y}_\alpha + zC_\alpha^\beta \mathbf{Y}_\beta, \quad (2.183)$$

from Eq. (2.131). It follows from Eq. (2.183) that

$$g_{\alpha\beta}(z) = \mathbf{Y}_\alpha(z) \cdot \mathbf{Y}_\beta(z) = g_{\alpha\beta}(1 - z^2 K) + 2z(1 + zH)C_{\alpha\beta} \quad (2.184)$$

and, therefrom, that

$$\sqrt{g(z)} = \sqrt{g}\left[1 + 2zH + z^2 K + \mathcal{O}(z^3)\right] \quad (2.185)$$

$$g^{\alpha\beta}(z) = g^{\alpha\beta}(1 - 3z^2 K) - 2z(1 - 3zH)C^{\alpha\beta} + \mathcal{O}(z^3) \quad (2.186)$$

$$\mathbf{Y}^\alpha(z) = (1 - z^2 K)\mathbf{Y}^\alpha - zC_\beta^\alpha \mathbf{Y}^\beta + 2z^2 H C_\beta^\alpha \mathbf{Y}^\beta + \mathcal{O}(z^3), \quad (2.187)$$

in which terms of order z^3 and higher have been dropped. Using these equations, we arrive at the useful relation,

$$\sqrt{g(z)}\,\mathbf{Y}^\alpha(z) = \sqrt{g}\left[(1 + z2H)\mathbf{Y}^\alpha - zC_\beta^\alpha \mathbf{Y}^\beta\right] + \mathcal{O}(z^3), \quad (2.188)$$

in which the coefficient of z^2 conveniently vanishes.

We are now in a position to calculate the net force on the small (multilayer) membrane patch defined by the coordinates (s_0^1, s_0^2) and $(s_0^1 + ds^1, s_0^2 + ds^2)$, as shown in Fig. 2.52(b). The net boundary force on this patch is the sum of contributions $\mathbf{F}_n$, $n = 1, 2, 3, 4$, from the four sides. Each of these forces can be evaluated both at the macroscopic level and at the microscopic level of the multilayer-membrane model. Thus, at the macroscopic level,

$$\mathbf{F}_1 = dl_1 \mathbf{T}(\hat{\mathbf{p}}_1) = dl_1 (\hat{\mathbf{p}}_1)_\alpha \mathbf{T}^\alpha$$
$$= \sqrt{g}\, ds^1 \mathbf{T}^2 = \sqrt{g}\, ds^1 \left(\mathbf{T}_\parallel^2 + T_\perp^2 \hat{\mathbf{n}}\right) = (\mathbf{F}_1)_\parallel + (\mathbf{F}_1)_\perp, \quad (2.189)$$

where we have used the relations $dl_1 = ds^1 \sqrt{g_{11}}$ and $\hat{\mathbf{p}}_1 = \sqrt{g/g_{11}}\, \mathbf{Y}^2$. On the other hand, the in-plane part of $\mathbf{F}_1$ can be calculated by adding up the tensions from the individual layers of the model membrane,

$$(\mathbf{F}_1)_\parallel = \int dz\, dl_1(z) t(z) \hat{\mathbf{p}}_1(z) = ds^1 \int dz\, t(z) \sqrt{g(z)}\, \mathbf{Y}^2(z)$$
$$= ds^1 \sqrt{g} \int dz\, t(z) \left[(1 + zH)\mathbf{Y}^2 - zC_\beta^2 \mathbf{Y}^\beta\right], \quad (2.190)$$

where we have used Eq. (2.188) for the last line. It is convenient to define the integrals (see Eq. (2.182)),

$$T_0 = T_0(s^1, s^2) \equiv \int dz\, t(z) \text{ and } \mathcal{M} = \mathcal{M}(s^1, s^2) \equiv \int dz\, zt(z), \quad (2.191)$$

corresponding to the net densities of force and bending moment across the membrane. This allows us to write Eq. (2.190) in the form,

$$(\mathbf{F}_1)_\parallel = ds^1 \sqrt{g} \left[(T_0 + 2H\mathcal{M})\mathbf{Y}^2 - \mathcal{M}C_\beta^2 \mathbf{Y}^\beta\right]. \quad (2.192)$$

Comparison of Eqs. (2.189) and (2.192) now permits $\mathbf{T}_\parallel^2$ to be written in terms of the quantities T_0 and $\mathcal{M}$. The same calculation may be done for the force $\mathbf{F}_2$. The overall result is the identification,

$$\mathbf{T}_\parallel^\alpha = (T_0 + 2H\mathcal{M})\mathbf{Y}^\alpha - \mathcal{M}C_\beta^\alpha \mathbf{Y}^\beta. \quad (2.193)$$

The definition,

$$\tau_0 \equiv T_0 + 2H\mathcal{M} - \frac{\kappa_b}{2}(2H - C_0)^2, \quad (2.194)$$

leads finally to

$$\mathbf{T}_\parallel^\alpha = \left(\tau_0 + \frac{\kappa_b}{2}(2H - C_0)^2\right)\mathbf{Y}^\alpha - \mathcal{M}C_\beta^\alpha \mathbf{Y}^\beta, \quad (2.195)$$

which is equivalent to the Helfrich stress tensor, Eq. (2.64), but with the normal component $\mathbf{T}_\perp^\alpha$ still undetermined.

At this point in the argument it is still unclear both why τ_0 as defined by Eq. (2.194) is uniform (Eq. (2.67))[50] and why the normal component

---

50) Indeed, the definition Eq. (2.194) appears quite arbitrary at this point. Its only significance lies in the fact that Eq. (2.67) requires that this particular combination of terms be spatially uniform.

of the stress tensor is given by $\mathbf{T}_\perp^\alpha = \partial_\alpha \mathcal{M}$ (Eq. (2.63)). Both these relations are a consequence of the conditions of mechanical equilibrium for the patch. The arguments are given in general form in the text[51] at Eqs. (2.49) (leading to Eq. (2.53) and thereby to Eq. (2.67)) and Eq. (2.57) (leading to Eq. (2.63)), respectively. It may be of pedagogical value to rephrase these arguments in terms of the equilibrium of the infinitesimal patch shown in Fig. 2.52. We carry through the calculation for the force equilibrium only, leaving the corresponding torque equilibrium as an exercise for the reader. Equation (2.189) gave the expression $\mathbf{F}_1 = ds^1 \sqrt{g}\, \mathbf{T}^2$, which is to be evaluated at $s^2 = s_0^2 + ds^2$. A similar treatment for $\mathbf{F}_3$ gives the same expression but with a minus sign, which is to be evaluated at $s^2 = s_0^2$. Thus, $\mathbf{F}_1 + \mathbf{F}_3 = ds^1 ds^2\, \partial_2(\sqrt{g}\, \mathbf{T}^2)$. Including the contributions from the 2 and 4 edges and adding in the pressure force gives for translational equilibrium of the patch,

$$\mathbf{F}_{\text{net}} = \Delta P \hat{\mathbf{n}} + \sum_{n=1}^{4} \mathbf{F}_n = ds^1 ds^2\, \partial_\alpha(\sqrt{g}\mathbf{T}^\alpha)$$

$$= dA\, (\Delta P \hat{\mathbf{n}} + D_\alpha \mathbf{T}^\alpha) = \mathbf{0}, \qquad (2.196)$$

where in writing the last equality, we have made use of Eq. (2.146). Equation (2.196) rederives Eq. (2.53). In evaluating Eq. (2.53) to reach Eq. (2.67), it is useful to use the chain-rule property of the covariant derivative, so

$$D_\alpha \mathbf{T}^\alpha = D_\alpha \big(T^{\alpha\beta} \mathbf{Y}_\beta + T_\perp^\alpha \hat{\mathbf{n}}\big)$$
$$= \big(D_\alpha T^{\alpha\beta}\big)\mathbf{Y}_\beta + T^{\alpha\beta}\big(D_\alpha \mathbf{Y}_\beta\big) + \big(D_\alpha T_\perp^\alpha\big)\hat{\mathbf{n}} + T_\perp^\alpha\big(D_\alpha \hat{\mathbf{n}}\big). \quad (2.197)$$

Collecting normal and in-plane components of Eq. (2.53) then leads to

$$\Delta P - T^{\alpha\beta} C_{\alpha\beta} + D_\alpha T_\perp^\alpha = 0 \text{ and}$$
$$D_\alpha T_{\alpha\beta} + T_\perp^\alpha C_\alpha^\beta = 0. \qquad (2.198)$$

Substituting the Helfrich stress, Eq. (2.64), and making liberal use of the differential relations Eqs. (2.149)–(2.154) from Appendix C leads to the Ou-Yang equations, Eq. (2.67), for membrane-shape equilibrium. These equations may, of course, be derived directly from the free energy functional using the variational approach(Capovilla and Guven 2002; Lomholt and Miao 2006; Fournier 2007), as discussed in Appendix C.4.

[51] It is, of course, possible to calculate directly the intrinsic moment for the layer model, $\mathbf{N}_{\text{int}} = \int dl\, dz\, (z\hat{\mathbf{n}}) \times \big(t(z)\hat{\mathbf{p}}(z)\big) \to \int dl\, \hat{\mathbf{n}} \times \hat{\mathbf{p}} \int dz\, zt(z)$ at lowest order in z, so that the argument leading to Eq. (2.63) gives directly $T_\perp^\alpha = \partial^\beta \int dz\, zt(z)$.

D.3
Inclusion of Gaussian Curvature

The simple fluid-membrane model described in Appendix D.3 does not give rise to Gaussian curvature effects. If a Gaussian curvature term of the form of Eq. (2.15) is present, generalization is required of the restricted treatment of torques in Section 2.4.2 at Eq. (2.62) and the following material. Thus, generally, the intrinsic boundary torque has the form,

$$\mathbf{N}_{\mathrm{int}} = \int_{\partial \Sigma} dl\, \mathbf{N}(\hat{\mathbf{p}}) = \int_{\partial \Sigma} dl\, p_\alpha \mathbf{N}^\alpha = \int_{\partial \Sigma} dl\, p_\alpha \left(N^{\alpha\beta} \mathbf{Y}_\beta + N^\alpha_\perp \hat{\mathbf{n}} \right), \quad (2.199)$$

where $\mathbf{N}(\hat{\mathbf{p}})$ is the linear density of boundary torque and the representation in terms of the direction-independent components $\mathbf{N}^\alpha$ follows from a calculation in full parallel to that given for $\mathbf{T}^\alpha$ in Appendix D.1. What $\mathbf{N}(\hat{\mathbf{p}})$ represents physically is the torque per unit edge length produced by the variation of the 3D material stress across the thickness of the membrane. Such a torque can have no normal component, so $N^\alpha_\perp$ vanishes identically.[52] Thus, by use of Gauss theorem,

$$\mathbf{N}_{\mathrm{int}} = \int_\Sigma dA\, D_\alpha \left(N^{\alpha\beta} \mathbf{Y}_\beta \right) = \int_\Sigma dA \left[\left(D_\alpha N^{\alpha\beta} \right) \mathbf{Y}_\beta - N^{\alpha\beta} C_{\alpha\beta} \hat{\mathbf{n}} \right], \quad (2.200)$$

so torque equilibrium, $\mathbf{N}_{\mathrm{net}} = \mathbf{0}$ (Eq. (2.60)), now requires,

$$N^{\alpha\beta} C_{\alpha\beta} = 0 \quad \text{and} \quad (2.201)$$

$$D_\alpha N^{\alpha\beta} = T^\alpha_\perp \epsilon_{\alpha\gamma} g^{\gamma\beta}, \quad (2.202)$$

where we have assumed that the stress tensor is symmetric. Equation (2.202) is equivalent to

$$T^\alpha_\perp = \epsilon^{\alpha\beta} g_{\beta\gamma} D_\sigma N^{\sigma\gamma}. \quad (2.203)$$

Equation (2.201) places a constraint on the form of $N_{\alpha\beta}$. The simple Helfrich form, Eq. (2.62), for the torque density gives $\mathbf{N}(\hat{\mathbf{p}}) = \mathcal{M}\hat{\mathbf{n}} \times \hat{\mathbf{p}} = \mathcal{M} p_\alpha \epsilon^{\alpha\beta} \mathbf{Y}_\beta$, so $N^{\alpha\beta}_{\mathrm{Helfrich}} = \mathcal{M}\epsilon^{\alpha\beta}$, from which it is easy to see that Eq. (2.201) is satisfied and Eq. (2.203) reduces to Eq. (2.63).

What happens when the Gaussian contribution, Eq. (2.15), is added to the Helfrich model? Recall that the Gauss–Bonnet theorem guarantees that this addition cannot in any way affect the equilibrium shapes of closed vesicles of fixed topology. It can be shown by variational methods (Capovilla and Guven 2002; Lomholt and Miao 2006; Fournier 2007) that there is no change in

---

[52] The argument depends only on the fact that the edge is assumed to be cut normally to the tangent plane, so that the normal vector $\hat{\mathbf{n}}$ is parallel to the exposed edge at each point of the boundary. The net torque due to the strip of edge dl is $dl \int dz (\mathbf{R} + z\hat{\mathbf{n}}) \times \mathbf{t}(z, \hat{\mathbf{p}})$, where $\mathbf{t}(z, \hat{\mathbf{p}})$ is the force density profile. Note that $\int dz\, \mathbf{t}(z, \hat{\mathbf{p}}) \equiv \mathbf{T}(\hat{\mathbf{p}})$, so the first term is already included in Eq. (2.58) and it is only the second term which is part of $\mathbf{N}_{\mathrm{int}}$.

form of the stress tensor $\mathbf{T}^\alpha$, that is, Eq. (2.64) remains valid. However, the torque-density tensor acquires an additional term,[53]

$$N^{\alpha\beta} = \mathcal{M}\epsilon^{\alpha\beta} + \kappa_g \left(2H\delta^\alpha_\gamma - C^\alpha_\gamma\right)\epsilon^{\gamma\beta}. \qquad (2.204)$$

Note that the extra term continues to satisfy Eq. (2.201). At first sight these two statements might appear inconsistent, since Eq. (2.203) suggests that a change in $N_{\alpha\beta}$ must produce a corresponding change in $T^\alpha_\perp$. What avoids this inconsistency is a property of the added term in Eq. (2.204): $D_\alpha(C^\gamma_\gamma \epsilon^{\alpha\beta} - C^\alpha_\delta \epsilon^{\delta\beta}) = 0$ by virtue of Eqs. (2.152) and (2.153). The upshot is that the Gaussian curvature term does not affect the shape equations, which only involve the stress tensor. The only way the Gaussian term can affect membrane shapes is via membrane boundaries, at which the response to applied torques does depend on κ_g via Eq. (2.204). It follows that it is only possible to infer values of κ_g from equilibrium membrane-shape observations in the presence of open boundaries.[54]

The simple model used in Appendix D.2 of isotropic layers with a distribution $t(z)$ of in-plane tensions cannot produce Gaussian-curvature effects. To incorporate such effects in a layer model, it is necessary to allow in-plane anisotropy and to write for the layer stress tensor density $\mathbf{t}^\alpha(z) = t^\alpha_\beta(z)\mathbf{Y}^\beta(z) + t^\alpha_\perp(z)\hat{\mathbf{n}}$. As discussed earlier, since the molecular constituents of the membrane transfer stress over distances comparable to membrane thickness, the z-dependent stresses induced by membrane bending cannot be relieved by fluid motion in individual layers. Thus, for anisotropic bending $C_1 \neq C_2$, in general we might expect t^α_β to be anisotropic. Since t^α_β is isotropic for a flat membrane, to first order in curvature we may write,

$$t^\alpha_\beta(z) = \left[t_0(z) + (2H - C_0)t_1(z)\right]\delta^\alpha_\beta + \mu_1(z)\left[2H\delta^\alpha_\beta - C^\alpha_\beta(z)\right], \qquad (2.205)$$

where the first term is the previous isotropic part $t(z)$, Eq. (2.181), and the second term represents a new anisotropic term generically comparable in magnitude to the $t_1(z)$ term of the isotropic part. It is straightforward to show that the anisotropic term produces the new term in Eq. (2.204) for the intrinsic torque density with the microscopic prescription $\kappa_g = \int dz\, z\mu(z)$. The only problem is that this same term also produces additional contributions in the stress tensor $\mathbf{T}^\alpha$ which, if present, would violate the condition (arising from the Gauss–Bonnet theorem) that a Gaussian term in the en-

[53] $\mathbf{N}_{\text{Helfrich}}(\hat{\mathbf{p}})$ has a magnitude which is independent of $\hat{\mathbf{p}}$ and always points along the edge. Both of these isotropy properties are compromised by the added term in Eq. (2.204). Note, however, that when the two principal curvatures are equal (so $\mathbf{C}$ is proportional to the unit matrix), then $N^{\alpha\beta} \sim \epsilon^{\alpha\beta}$, so $\hat{\mathbf{p}} \cdot \mathbf{N}(\hat{\mathbf{p}}) = 0$ and isotropy is restored. This conditional isotropy reflects membrane fluidity.

[54] This statement assumes that topology is fixed, as it is for highly insoluble lipids. By contrast, for systems like microemulsions, in which topology change via the solution is allowed on experimental time scales, the Gaussian modulus can affect the equilibrium distribution of shapes and is observable.

ergy cannot influence the membrane stresses. Thus, it is not hard to repeat the calculations, Eqs. (2.189)–(2.193),

$$\mathbf{T}_{\parallel}^{\alpha} = \int dz\, t_{\beta}^{\alpha}(z) \sqrt{\frac{g(z)}{g}} \mathbf{Y}^{\beta}$$

$$= \int dz\, t_{\beta}^{\alpha}(z) \left[(1 + 2zH)\mathbf{Y}^{\beta} - zC_{\gamma}^{\beta}\mathbf{Y}^{\gamma} + \mathcal{O}\left(z^{3}\right) \right]. \qquad (2.206)$$

At leading order, the anisotropic part of the stress tensor generates a new term of order z (membrane thickness),

$$\left(\int dz\, \mu_{1}(z) \right) \left(2H\delta_{\beta}^{\alpha} - C_{\beta}^{\alpha} \right) \mathbf{Y}^{\beta}. \qquad (2.207)$$

Thus, the condition $\int dz\, \mu_{1}(z) = 0$ must be satisfied in order that the stress tensor be independent of μ_{1}. This condition, which must arise from the microscopic mechanics, is presumably related to 3D membrane fluidity. Finally, at order z^{2}, we find,

$$\Delta \mathbf{T}_{\parallel}^{\alpha} = 4H\kappa_{g}\left(H\delta_{\beta}^{\alpha} - C_{\beta}^{\alpha}\right)\mathbf{Y}^{\beta}. \qquad (2.208)$$

The cancellation of these terms requires counter-terms of order z^{2} in Eq. (2.205). The physical origin of these cancellations is buried in the microscopic material mechanics, which is beyond the scope of this discussion.

D.4
Deformation Matrix M in Curvilinear Coordinates

We next consider membrane mechanics for the composite membrane consisting of the fluid plasma membrane coupled to the membrane skeleton. To treat the mechanics of the membrane skeleton in a more general way than was done in Section 2.4.3, we will need to be able to express the local deformation matrix **M** introduced in Appendix B in a properly covariant manner. **M** expresses the linear deformation of a small (infinitesimal) local patch of membrane near the arbitrary point (s_0^1, s_0^2) during the transformation of the cytoskeleton from its undeformed configuration $\mathbf{R}_0$ to its final deformed configuration $\mathbf{R}$. It will be convenient to treat this neighborhood as (locally) flat and to write

$$\mathbf{R}_0\left(s^1, s^2\right) - \mathbf{R}_0\left(s_0^1, s_0^2\right) = \mathbf{X}_0 = a_i\left(s^1, s^2\right) \hat{\mathbf{a}}_i\left(s_0^1, s_0^2\right), \qquad (2.209)$$

where the two Cartesian unit vectors $\hat{\mathbf{a}}_i$ define the local tangent plane of the undeformed shape $\mathbf{R_0}$ at (s_0^1, s_0^2). A similar representation of the deformed patch is given by

$$\mathbf{R}\left(s^1, s^2\right) - \mathbf{R}\left(s_0^1, s_0^2\right) = \mathbf{X} = x_i\left(s^1, s^2\right) \hat{\mathbf{x}}_i\left(s_0^1, s_0^2\right). \qquad (2.210)$$

Note that the 3D planes defined by the unit vectors $\hat{\mathbf{a}}_i$ and $\hat{\mathbf{x}}_i$ are generally different and that the choices of Cartesian axes in these planes are arbitrary and unrelated. The covariant tangent vectors are given by (Eq. 2.119)

$$\mathbf{Y}_\alpha = \frac{\partial \mathbf{R}}{\partial s^\alpha} = \frac{\partial x_i}{\partial s^\alpha}\hat{\mathbf{x}}_i \equiv Y_{\alpha i}\hat{\mathbf{x}}_i, \quad \text{so } \hat{\mathbf{x}}_i = Y_i^\alpha \mathbf{Y}_\alpha. \tag{2.211}$$

The corresponding contravarient tangent vectors must obey $\mathbf{Y}_\alpha \cdot \mathbf{Y}^\beta = \delta_\alpha^\beta$ (Eq. (2.125)), so we may conclude

$$Y_{\alpha i} = \frac{\partial x_i}{\partial s^\alpha} \quad \text{and} \quad Y_i^\alpha = \frac{\partial s^\alpha}{\partial x_i}, \tag{2.212}$$

and similarly for the tangent vectors $\mathbf{Y}_{0\alpha}$ and $\mathbf{Y}_0^\alpha$ of the undeformed shape. Orthogonality and completeness of the tangent vectors lead to the useful relations

$$Y_{\alpha i} Y_i^\beta = \delta_\alpha^\beta \quad \text{and} \quad Y_{\alpha i} Y_j^\alpha = \delta_{ij}. \tag{2.213}$$

The relation $x_i = M_{ij} a_j$ which defines the deformation matrix connects the Cartesian coordinates of points of $\mathbf{X}$ and $\mathbf{X}_0$ which map into one another, that is, points with the same labels (s^1, s^2). It follows that

$$M_{ij} = \frac{\partial x_i}{\partial a_j}\bigg|_s = \frac{\partial x_i}{\partial s^\alpha}\frac{\partial s^\alpha}{\partial a_j} = Y_{\alpha i} Y_{0j}^\alpha, \tag{2.214}$$

where we have used Eq. (2.212) at the last step. Note that the ordering of the lower/upper indices in Eq. (2.214) is significant, because raising the index of $\mathbf{Y}_\alpha$ involves the metric tensor g while lowering the index of $\mathbf{Y}_0{}^\alpha$ involves the metric tensor g_0. M_{ij} is explicitly dependent on the arbitrary choices of Cartesian axes $\hat{\mathbf{a}}_i$ and $\hat{\mathbf{x}}_i$; however, in the calculation of physical quantities this dependence will disappear. Thus, for example, it is easy to show from Eq. (2.214) that

$$\lambda_1 \lambda_2 = \det \mathbf{M} = |\mathbf{Y}_0^1 \times \mathbf{Y}_0^2| = \sqrt{\frac{g}{g_0}} = \frac{dA}{dA_0} \quad \text{and}$$

$$(\lambda_1^2 + \lambda_2^2) = \operatorname{tr} \mathbf{M}^T \mathbf{M} = \operatorname{tr} \mathbf{g}\,\mathbf{g}_0^{-1}, \tag{2.215}$$

which allow the area and shear strains Eqs. (2.118) to be expressed in terms of the metric tensors.

It is sometimes convenient to represent the matrix $\mathbf{M}$ covariantly, which can be done in in terms of either the undeformed or deformed bases. Thus, for example, $M_{ij} = Y_{\alpha i} M^\alpha{}_\beta Y_j^\beta$ with $M^\alpha{}_\beta = Y_{0j}^\alpha Y_{\beta j}$. Functions of $\mathbf{M}$ like the Cauchy strain tensor Eq. (2.113) may also be expressed in terms of the tangent vectors. For example, direct calculation shows that

$$u_{ij} = \frac{1}{2}\left(Y_{0i}^\alpha g_{\alpha\beta} Y_{0j}^\beta - \delta_{ij}\right), \tag{2.216}$$

where **g** is the metric tensor of the deformed surface. Equivalently, in the undeformed representation, one finds

$$u_{\alpha\beta} \equiv Y_{0\alpha i} u_{ij} Y_{0\beta j} = \frac{1}{2}\left[g_{\alpha\beta} - (g_0)_{\alpha\beta}\right], \qquad (2.217)$$

which generalizes Eq. (2.114) to curvilinear coordinates in that $2u_{\alpha\beta} ds^\alpha ds^\beta = dl^2 - dl_0^2$.

D.5
Stress Tensor for the Membrane Skeleton

The stress tensor $\mathbf{T}_{\text{ms}}^{\alpha\beta}$ of the membrane skeleton is needed in order to compute the normal force Q, Eq. (2.76), which acts between the plasma membrane and the membrane skeleton and, thus, to carry through the program outlined in Section 2.4.3 to find the equilibrium shape of the composite membrane. The so-called constitutive relations which connect the stresses $\mathbf{T}_{\text{ms}}^{\alpha\beta}$ to the strains α and β, Eqs. (2.24) and (2.25), or, equivalently, to the local elastic deformation matrix **M** (Appendix B.2) are implicit in Eq. (2.28) for the elastic energy F_{ms} or in the expression for the local elastic energy density $f_{\text{ms}}(\mathbf{M})$ (Appendix B.2). In this section we show how to derive Eq. (2.80) for $\mathbf{T}_{\text{ms}}^{\alpha\beta}$ which expresses this connection. In the remainder of this section we will drop the subscript "ms" on the stress tensor to simplify the notation.

Consider an infinitesimal membrane skeletal patch (Fig. 2.52(b)) defined by the coordinates (s_0^1, s_0^2) and $(s_0^1 + ds^1, s_0^2 + ds^2)$. The change δF in the elastic energy of this patch under a change of mapping from $\mathbf{X}_0 \to \mathbf{X}$ to $\mathbf{X}_0 \to \mathbf{X} + \delta\mathbf{X}$ (in the notation of Eqs. (2.209) and (2.210)) can be calculated in two different ways. On the one hand,

$$\delta F = dA_0 \frac{\partial f_{\text{ms}}}{\partial M_{ij}} \delta M_{ij}, \qquad (2.218)$$

where $\delta M_{ij} = \delta(Y_{\alpha i} Y_{0j}^\alpha) = Y_{0j}^\alpha \delta Y_{\alpha i} = Y_{0j}^\alpha \left[\hat{\mathbf{x}}_i \cdot \frac{\partial(\delta\mathbf{R})}{\partial s^\alpha}\right]$. On the other hand, this change in energy must be equal to the work done by the boundary forces acting on the deformed patch as it moves from its position under the deformation $\mathbf{X}$ to its position under $\mathbf{X} + \delta\mathbf{X}$. This work may be calculated by adding up contributions, δW_n, from the four boundaries:

$$\delta W = \sum_{n=1}^{4} \delta W_n = \mathbf{F}_n \cdot \delta\mathbf{R}_n = \sum_{n=1}^{4} dl_n \mathbf{T}(\hat{\mathbf{p}}_n) \cdot \delta\mathbf{R}_n, \qquad (2.219)$$

where $\mathbf{F}_n$ is the force acting at the midpoint of side n and $\delta\mathbf{R}_n$ is the corresponding displacement under $\delta\mathbf{X}$. It will be convenient to define the stress tensor,[55]

---

55) T_{ij} may be viewed as the stress tensor in the local Cartesian coordinates defined by the unit vectors $\hat{\mathbf{x}}_1, \hat{\mathbf{x}}_2$ of Appendix D.4.

$$T_{ij} \equiv Y_{\alpha i} T^{\alpha\beta} Y_{\beta j}, \text{ so } T^{\alpha\beta} = Y_i^\alpha T_{ij} Y_j^\beta. \tag{2.220}$$

In this notation, for example, $\mathbf{F}_1 = dl_1 \mathbf{T}(\hat{\mathbf{p}}_1) = \sqrt{g_{11}}\, ds^1 (\hat{\mathbf{p}}_1)_\alpha T^{\alpha\beta} \mathbf{Y}_\beta = \sqrt{g}\, ds^1 Y_i^2 T_{ij} \hat{\mathbf{x}}_j$, where we have used $\hat{\mathbf{p}}_1 = \sqrt{g/g_{11}}\, \mathbf{Y}^2$ and Eq. (2.211). Combining the contributions from sides 1 and 3 gives (to lowest order in ds^1, ds^2)

$$\delta W_1 + \delta W_3 = \sqrt{g}\, ds^1 Y_i^2 T_{ij} \hat{\mathbf{x}}_j \cdot (\delta \mathbf{R}_1 - \delta \mathbf{R}_3) = dA\, T_{ij} Y_i^2 \hat{\mathbf{x}}_j \cdot \frac{\partial(\delta \mathbf{R})}{\partial s^2}, \tag{2.221}$$

where we have used $dA = \sqrt{g}\, ds^1 ds^2$ and $\delta \mathbf{R}(s_0^1, s_0^2) = 0$. Adding in the contributions from sides 2 and 4, we obtain

$$\delta W = dA\, T_{ij} Y_i^\alpha \hat{\mathbf{x}}_j \cdot \frac{\partial(\delta \mathbf{R})}{\partial s^\alpha}. \tag{2.222}$$

Equating δW and δF and noting $dA/dA_0 = \det \mathbf{M}$ (Eq. (2.215)), we arrive finally at the key formula,

$$T_{ij} = \frac{1}{\det \mathbf{M}} M_{ik} \left(\frac{\partial f_{\mathrm{ms}}}{\partial M_{jk}} \right). \tag{2.223}$$

To evaluate Eq. (2.223) we need the chain rule, $\frac{\partial f_{\mathrm{ms}}}{\partial M_{jk}} = \frac{\partial f_{\mathrm{ms}}}{\partial \alpha} \frac{\partial \alpha}{\partial M_{jk}} + \frac{\partial f_{\mathrm{ms}}}{\partial \beta} \frac{\partial \beta}{\partial M_{jk}}$, plus Eq. (2.118) and the identities, $M_{ik} \frac{\partial(\det \mathbf{M})}{\partial M_{jk}} = \det \mathbf{M}\, \delta_{ij}$ and $\frac{\partial(\mathrm{tr}\, \mathbf{M}^T \mathbf{M})}{\partial M_{jk}} = 2 M_{jk}$. The result is

$$T_{ij} = \left(\frac{\partial f_{\mathrm{ms}}}{\partial \alpha} \right) \delta_{ij} + \frac{1}{(\det \mathbf{M})^2} \left(\frac{\partial f_{\mathrm{ms}}}{\partial \beta} \right) \left[M_{ik} M_{jk} - \mathrm{tr}\left(\mathbf{M}^T \mathbf{M} \right) \delta_{ij}/2 \right], \tag{2.224}$$

from which we find from Eq. (2.220)

$$T^{\alpha\beta} = \left(\frac{\partial f_{\mathrm{ms}}}{\partial \alpha} \right) g^{\alpha\beta} + \frac{1}{(1+\alpha)^2} \left(\frac{\partial f_{\mathrm{ms}}}{\partial \beta} \right) \left[g_0^{\alpha\beta} - (1+\alpha)(1+\beta) g^{\alpha\beta} \right], \tag{2.225}$$

which is given in the text as Eq. (2.80).

D.6
Shape Equations Under Conditions of Axisymmetry: Some Examples

We will derive in this section explicit shape equations for fluid membranes with bending rigidity in axisymmetric geometries. The parametrization, illustrated in Fig. 2.12, involves as surface coordinates the arc length s along a line of longitude and the azimuthal angle ϕ. The radial distance from the symmetry axis is r and θ denotes the angle of the surface away from the radial direction, so that

$$\frac{dr}{ds} = \cos \theta. \tag{2.226}$$

In the general curvilinear notation of Appendix C, we take $(s^1, s^2) \equiv (s, \phi)$, so $\mathbf{Y}_1 = \hat{\mathbf{s}}$ and $\mathbf{Y}_2 = r\hat{\boldsymbol{\phi}}$. It then follows that

$$\mathbf{g} = g_{\alpha\beta} = \begin{bmatrix} 1 & 0 \\ 0 & r^2 \end{bmatrix} \quad \text{with} \quad g = r^2 \quad \text{and}$$

$$\mathbf{C} = C^\alpha_\beta = \begin{bmatrix} C_m & 0 \\ 0 & C_p \end{bmatrix} \quad \text{with} \quad C_m = \frac{d\theta}{ds} = \frac{d\sin\theta}{dr}, \; C_p = \frac{\sin\theta}{r}, \quad (2.227)$$

so $2H = \text{tr}\mathbf{C} = C_m + C_p$ and $\det\mathbf{C} = C_m C_p$. The only non-vanishing Christoffel symbols are $\Gamma^2_{12} = \Gamma^2_{21} = \frac{1}{r}\frac{dr}{ds}$ and $\Gamma^1_{22} = -r\frac{dr}{ds}$. Because of axisymmetry, all these quantities depend on s alone and are independent of ϕ. It follows from Eq. (2.227) that

$$\frac{dC_p}{dr} = \frac{1}{r}(C_m - C_p), \quad (2.228)$$

which is, again, a consequence of axisymmetry.

First, we consider the simple fluid membrane with bending rigidity, as treated in Section 2.4.2, where we derived two forms of the shape equations, the general Ou-Yang equations, Eq. (2.67), based on the stress-tensor analysis, and the more-restrictive Helfrich equation, Eq. (2.69), based on axial force balance in the axisymmetric geometry. The Ou-Yang equation, which is generally a partial differential equation in the two surface variables (s^1, s^2), becomes for the axisymmetric geometry an ordinary differential equation in the arc length variable s, since $2H = C_m(s) + C_p(s)$ and

$$\Delta\mathcal{M} = \frac{1}{\sqrt{g}}\frac{d}{ds}\left(\sqrt{g}\frac{d(2H)}{ds}\right) = \frac{\kappa_b}{r}\frac{d}{ds}\left(r\frac{d(C_m + C_p)}{ds}\right). \quad (2.229)$$

With this relation substituted, Eq. (2.67) along with Eq. (2.226) and the definitions Eq. (2.227) of C_m and C_p become a system of four ordinary differential equations for the four variables $C_m(s), C_p(s), r(s)$, and $\theta(s)$ which can be solved simultaneously for the equilibrium shape. The Helfrich equation (Deuling and Helfrich 1976) which, in the notation of the present section takes the form,

$$\kappa_b \frac{d}{dr}(C_m + C_p) =$$

$$\frac{1}{[1 - (rC_p)^2]}\left[\kappa_b\left((C_p - C_0)^2 - C_m^2\right)C_p + 2\tau_0 C_p - \Delta P\right], \quad (2.230)$$

constitutes a first integral of this system under conditions such that the axial tension vanishes (e.g., for a closed vesicle).

Next, we derive shape equations for an axisymmetric composite membrane consisting of a fluid membrane of the Helfrich type described above[56]

---

56) It may be useful to remind the reader here that, in both Eq. (2.230) above and what follows below for the composite membrane, the area-difference elasticity Eq. (2.16) is still absent. To include this effect requires the further replacement of C_0 by C_0^{eff}, Eq. (2.41).

but now associated with an elastic membrane skeleton of the type described in Section 2.4.3 which is also axisymmetric. For simplicity, we will imagine that the unstrained shape S_0 of the membrane skeleton is a planar disk, whose center is coincident with the pole P of Fig. 2.12 and which is then deformed in an axisymmetric manner to fit onto the patch and allowed to come to elastic equilibrium. Such a geometry is relevant, for example, to calculation of the shape of echinocytic spicules and has been discussed in this context by Mukhopadhyay et al. (2002). Suppose that the radial variable of the undeformed patch is denoted s_0 with $s = s_0 = 0$ at the pole P. Then, the elastic state of the membrane skeleton when it is deformed to conform to the patch is completely specified by giving the mapping $s(s_0)$ (or equivalently $s_0(s)$) with fixed $\phi = \phi_0$. The principal axes of the strain are now the coordinate axes (s, ϕ), and it is easy to see that the principal extension ratios (Section 2.3.3 and Appendix B) take the form,

$$\lambda_1 = \frac{ds}{ds_0} \quad \text{and} \quad \lambda_2 = \frac{r}{s_0}, \tag{2.231}$$

from which the strain invariants α and β, Eqs. (2.24)–(2.25), can be calculated. The stress tensor can be calculated from Eq. (2.225) (Eq. (2.80)) and turns out to be diagonal in the (s, ϕ) representation by virtue of the axisymmetry, $\mathbf{T}_{ms} = (T_{ms})^\alpha_\beta = \begin{bmatrix} \tau_1 & 0 \\ 0 & \tau_2 \end{bmatrix}$ with,

$$\tau_1 = \frac{\partial f_{ms}}{\partial \alpha} + \frac{\partial f_{ms}}{\partial \beta} \frac{1}{2} \left(\frac{1}{\lambda_2^2} - \frac{1}{\lambda_1^2} \right)$$

and

$$\tau_2 = \frac{\partial f_{ms}}{\partial \alpha} + \frac{\partial f_{ms}}{\partial \beta} \frac{1}{2} \left(\frac{1}{\lambda_1^2} - \frac{1}{\lambda_2^2} \right), \tag{2.232}$$

which are equivalent to Eq. (2.79). Note that Eq. (2.231) allows $\lambda_1, \lambda_2, \alpha, \beta, \tau_1$ and τ_2 all to be expressed in terms of the two unknown functions $s_0(s)$ and $r(s)$.

The general equations for equilibrium of the membrane skeleton have been derived in Section 2.4.3 as Eqs. (2.76)–(2.77). In the axisymmetric case, these equations become, respectively,

$$Q = \tau_1 C_m + \tau_2 C_p$$
$$= \frac{\partial f_{ms}}{\partial \alpha}(C_m + C_p) - \frac{\partial f_{ms}}{\partial \beta} \frac{1}{2}\left(\frac{1}{\lambda_1^2} - \frac{1}{\lambda_2^2}\right)(C_m + C_p) \tag{2.233}$$

$$0 = D_\alpha T^\alpha_\beta \equiv \partial_\alpha T^\alpha_\beta + \Gamma^\alpha_{\alpha\gamma} T^\gamma_\beta - \Gamma^\gamma_{\alpha\beta} T^\alpha_\gamma$$
$$= \frac{d\tau_1}{ds} + (\tau_1 - \tau_2)\frac{1}{r}\frac{dr}{ds}, \tag{2.234}$$

in which we have inserted explicitly the Christoffel symbols from below Eq. (2.227). Equation (2.233) expresses the normal pressure Q on the cytoskeleton due to the plasma membrane (positive when pointing outward), whose negative is the corresponding pressure of the cytoskeleton on the plasma membrane. It is not hard to show that Eq. (2.234) expresses axial force balance for the infinitesimal slice of membrane skeleton between s to $s+ds$. To complete the treatment of the composite membrane, it is now necessary to incorporate the cytoskeletal tensions, Eqs. (2.232) or (2.233), into the membrane-shape equations, as discussed at the end of Section 2.4.3. This can be done in two ways. On the one hand, one can make the substitution $\Delta P \to \Delta P - Q$ in the Ou-Yang equation (2.67). Alternatively, one can add the cytoskeletal tension τ_1 to the axial force balance, Eq. (2.68), as $\mathbf{T}_{\hat{p}} \to \mathbf{T}_{\hat{p}} + \tau_1$.[57] This change implies the replacement $\tau_0 \to \tau_0 + \tau_1$ in Eqs. (2.69) and (2.230).

A full set of equations for the axisymmetric shape problem of the composite membrane requires five equations for the five unknown functions $s_0(s), C_m(s), C_p(s), r(s)$ and $\theta(s)$. Three of these are provided as previously by Eq. (2.226) and the expressions, Eqs. (2.227), for $C_{m,p}$. The fourth is Eq. (2.234). The final equation is the shape equation, either in the Ou-Yang form Eq. (2.67) with $\Delta P \to \Delta P - Q$ or in the (once-integrated) Helfrich form Eq. (2.230) with $\tau_0 \to \tau_0 + \tau_1$. By way of illustration, for linear elasticity (Eq. (2.27)), Eqs. (2.234) takes the particularly simple form,

$$\frac{d}{ds}\left[K_\alpha\left(\frac{r}{s_0}\frac{ds}{ds_0}-1\right) - \frac{\mu}{2}\left\{\left(\frac{ds_0}{ds}\right)^2 - \frac{s_0^2}{r^2}\right\}\right] = \frac{\mu}{r}\frac{dr}{ds}\left[\left(\frac{ds_0}{ds}\right)^2 - \frac{s_0^2}{r^2}\right], \tag{2.235}$$

while $\tau_0 \to \tau_0 + K_\alpha\left(\frac{r}{s_0}\frac{dr}{ds}-1\right) - \frac{\mu}{2}\left[\left(\frac{ds_0}{ds}\right)^2 - \frac{s_0^2}{r^2}\right]$ in the Helfrich equation (2.230).

---

57) Note that in this context Eq. (2.68) applies to the composite membrane, so the normal forces Q and $-Q$ between the plasma membrane and the membrane skeleton are internal forces.

References

Alberts, B., Johnson, A., Lewis, J., Raff, M.,Roberts, K., and Walter, P., 2002, *Molecular Biology of the Cell*. Garland Science, New York.

Allen, M. P., and Tildesley, D. J., 1989, *Metropolis Monte Carlo method*. Oxford, New York.

Artmann, G. M., Sung, K.-L. P.. Horn, T., Whittemore, D., Norwich, G., and Chien, S., 1997, *Biophys. J.* **72**, 1434.

Baba, T., Terada, N., Fujii, Y., Ohno, N.,Ohno, S., and Sato, S. B., 2004, *Histochem. Cell. Biol.* **122**, 587.

Başar, Y., and Weichert, D., 2000, *Nonlinear Continuum Mechanics of Solids: Fundamental Mathematical and Physical Concepts*. Springer-Verlag, Berlin.

Bennett, V., and Baines, A. J., 2001, *Physiol. Rev.* **81**, 1353.

Bessis, M., 1956, *Cytology of the Blood and Blood-Forming Organs*. Grune & Stratton, New York.

Bessis, M., 1972, *Nouv. Rev. Fr. Hematol.* **12**, 721.

Bessis, M., 1973, *Living Blood Cells and their Ultrastructure*, translated by R.I. Weed. Springer-Verlag, New York.

Bessis, M., 1974, *Corpuscles: Atlas of Red Blood Cell Shapes*. Springer-Verlag, Berlin.

Bessis, M., 2000, in *Hematol. Cell Ther.* **4**, 265–344 (Springer, 2000). This fourth and final issue of a discontinued journal, in which all of Bessis's papers are reprinted, is difficult to locate and not avavilable from the publisher's web site.

Betticher, D. C., Reinhart, W. H., and Geiser, J., 1995, *J. Appl. Physiol.* **78**, 778.

Blank, M. E., Hoefner, D. M., and Diedrich, D. F., 1994, *Biochim. Biophys. Acta* **1192**, 223.

Boal, D. H., 1994, *Biophys. J.* **67**, 521.

Boal, D. H., 2002, *Mechanics of the Cell*, ch. 3. Cambridge University Press, Cambridge.

Boal, D. H., Seifert, U., and Zilker, A., 1992, *Phys. Rev. Lett.* **69**, 3405.

Boon, J. M., and Smith, B. D., 2002, *Med. Res. Rev.* **22**, 251.

Brailsford, J. D., Korpman, R. A., and Bull, B. S., 1980, *J. Theor. Biol.* **86**, 531.

Brochard, F., and Lennon, J. F., 1975, *J. Phys. (France)* **36**, 1035.

Bukman, D. J., Yao, J. H., and Wortis, M., 1996, *Phys. Rev. E* **54**, 5463.

Canham, 1970, *J. Theor. Biol.* **26**, 61.

Canham, P. B., and Burton, A. C., 1968, *Circ. Res.* **22**, 405.

Canham, P. B., and Parkinson, D. R., 1970, *Can. J. Physiol. Pharmacol.* **48**, 369.

Capovilla, R., and Guven, J., 2002, *J. Phys. A: Math. Gen.* **35**, 6233.

Chaikin, P. M., and Lubensky, T. C., 1995, *Principles of Condensed Matter Physics*. Cambridge University Press, Cambridge.
Chi, L.-M., and Wu, W.-G., 1990, *Biophys. J.* **57**, 1225.
Cuvelier, D., Derényi, I., Bassereau, P., and Nassoy, P., 2005, *Biophys. J.* **88**, 2714.
Dao, M., Lim, C. T., and Suresh, S., 2003, *J. Mech. Phys. Solids* **51**, 2259.
Deuling, H. J., and Helfrich, W., 1976, *J. Phys. (Paris)* **37**, 1335.
Deuticke, B., 1968, *Biochim. Biophys. Acta* **163**, 494.
Deuticke, B., Grebe, R., and Haest, C. W. M., 1990, in *Blood Cell Biochemistry*, pp. 475–529. Plenum Press, New York.
Discher, D. E., Boal, D. H., and Boey, S. K., 1998, *Biophys. J.* **75**, 1584.
Discher, D. E., and Mohandas, N., 1996, *Biophys. J.* **71**, 1680.
Discher, D. E., Mohandas, N., and Evans, E. A., 1994, *Science* **266**, 1032.
do Carmo, M. P., 1976, *Differential Geometry of Curves and Surfaces*. Prentice-Hall, Engelwood Cliffs, NJ.
Elgsaeter, A., Stokke, B. T., Mikkelsen, A., and Branton, D., 1986, *Science* **234**, 1217.
Engström, K. G., and Löfvenberg, E., 1998, *Blood* **91**, 3986.
Evans, E., 1974, *Biophys. J.* **14**, 923.
Evans, E., 1980, *Biophys. J.* **30**, 265.
Evans, E. A., 1983, *Biophys. J.* **43**, 27.
Evans, E. A., and Fung, Y. C., 1972, *Microvasc. Res.* **4**, 335.
Evans, E, and Rawicz, W., 1990, *Phys. Rev. Lett.* **64**, 2094.
Evans, E. A., and Skalak, R., 1980, *Mechanics and Thermodynamics of Biomembranes*. CRC Press, Boca Raton, FL.
Ferrell, J. E. Jr., Lee, K.-J., and Huestis, W. H., 1985, *Biochemistry* **24**, 2849.
Fischer, T. M., 2004, *Biophys. J.* **86**, 3304.
Fischer, T. M., 2006, private communication.
Fischer, T. M., Haest, C. W. M., Stöhr-Liesen, M., Schmid-Schönbein, H., and Skalak, R. , 1981, *Biophys. J.* **34**, 409.
Fischer, T, M., and Schmidt-Schönbein, H., 1977, *Blood Cells* **3**, 351.
Fourcade, B., Miao, L., Rao, M., Wortis, M., and Zia, R. K. P., 1994, *Phys. Rev. E* **49**, 5276.
Fournier, J.-B., 2007, *Soft Matter* **3**, 883.
Fung, Y. C., and Tong, P., 1968, *Biophys. J.* **8**, 175.
Fung, Y. C., Tsang, W. C. O., and Patitucci, P., 1981, *Biorheology* **18**, 369.
Furchgott, R. F., and Ponder, E., 1940, *J. Exptl. Biol.* **17**, 117.
Gedde, M. M., Davis, D. K., and Huestis, W. H., 1997, *Biophys. J.* **72**, 1234.
Gedde, M. M., Yang, E., and Huestis, W. H., 1995, *Blood* **86**, 1595.
Gedde, M. M., Yang, E., and Huestis, W. H., 1999, *Biochim. Biophys. Acta* **1417**, 246.
Gennis, R. B., 1989, *Biomembranes*. Springer-Verlag, New York.

Gompper, G., and Kroll, D. M., 1997, *J. Phys. Condens. Matter* **9**, 8795.
Guck, J., Ananthakrishnan, R., Mahmood, H., Moon, T. J., Cunningham, C. C., and Käs, J., 2001, *Biophys. J.* **81**, 767.
Hamburger, H. J., 1895, *Pflügers Arch.* **141**, 230. This citation is given in Ponder (1948); however, we have been unable to verify it.
Hansen, J. C., Skalak, R., Chien, S., and Hoger, A., 1996, *Biophys. J.* **70**, 146.
Heinrich, V., Ritchie, K., Mohandas, N., and Evans, E., 2001, *Biophys. J.* **81**, 1452.
Helfrich, W., 1973, *Z. Naturforsch.* **28c**, 693.
Helfrich, W., 1974, *Z. Naturforsch.* **29c**, 510.
Hénon, S., Lenormand, G., Richert, A., and Gallet, F., 1999, *Biophys. J.* **76**, 1145.
Hochmuth, R. M., 1993, *J. Biomech. Eng.* **115**, 515.
Hochmuth, R. M., and Mohandas, N., 1972, *Microvasc. Res.* **4**, 295.
Hoefner, D. M., Blank, M. E., Davis, B. M., and Diedrich, D. F., 1994, *J. Membr. Biol.* **141**, 91.
Hoffman, J. F., 2001, *Blood Cells Mol. Dis.* **27**, 57.
Holzapfel, G. A., 2000, *Nonlinear Solid Mechanics: A Continuum Approach for Engineering.* Wiley, New York.
Humpert, C., and Baumann, M., 2003, *Mol. Membr. Biol.* **20**, 155.
Hwang, W. C., and Waugh, R. E., 1997, *Biophys. J.* **72**, 2669.
Iglič, A., 1997, *J. Biomech.* **30**, 35.
Iglič, A., Krajl-Iglič, V., and Hägerstrand, H., 1998, *Med. Biol. Eng. Comput.* **36**, 251.
Iglič, A., Krajl-Iglič, V., and Hägerstrand, H., 1998, *Eur. Biophys. J.* **27**, 335.
Isomaa, B., Hägerstrand, H., and Paatero, G., 1987, *Biochim. Biophys. Acta* **899**, 93.
Jacobson, K., Mouritsen, O. G., and Anderson, G. W., 2007, *Nature Cell Biol.* **9**, 7.
Jain, N. C., 1975, *Blood Cells* **1**, 385.
Jain, N. C., and Keeton, K. S., 1974, *Br. Vet. J.* **130**, 288.
Jain, N. C., and Kono, C. S., 1972, *Res. Vet. Sci.* **13**, 489.
Jay, A. W. L., 1975, *Biophys. J.* **15**, 205.
Johnson, R. M., Taylor, G., and Meyer, D. B., 1980, *J. Cell. Biol.* **86**, 371.
Jones, B., Walker, T. F., Chahwala, S. B., Thompson, M. G., and Hickman, J. A., 1987, *Exp. Cell Res.* **168**, 309.
Jülicher, F., 1996, *J. Phys. II* **6**, 1797.
Kantor, Y., 1989, in *Statistical Mechanics of Membranes and Surfaces*, edited by D. Nelson, T. Piran, and S. Weinberg, Vol. 5 of *Jerusalem Winter School for Theoretical Phsics*, pp. 115–136. World Scientific, Singapore.
Kantor, Y., and Nelson, D. R., 1987, *Phys. Rev. Lett.* **58**, 2774.
Katnik, C. and Waugh, R. E., 1990, *Biophys. J.* **57**, 877.

Kern, N., 1998, PhD thesis, Université Joseph Fournier, Grenoble, France (unpublished).
Khairy, K., Foo, J., and Howard, J., 2007 (unpublished).
Khodadad, J. K., Waugh, R. E., Podolski, J. L., Josephs, R., and Steck, T. L., 1996, *Biophys. J* **70**, 1036.
Kimzey, S. L., 1977, in *Biomedical Results from Skylab*, edited by R. S. Johnston and L. F. Dietlein, pp. 249–282. National Aeronautics and Space Administration, Washington, DC.
Kreyszig, E., 1991, *Differential Geometry*. Dover Publications Inc., New York.
Kuzman, D., Svetina, S., Waugh, R. E., and Žekš, B., 2004, *Eur. Biophys. J.* **33**, 1.
Lang, F., 2006, *Mechanisms and Significance of Cell Volume Regulation*. Karger, Basel.
Lange, Y., Hadesman, R. A., and Steck, T. L., 1982, *J. Cell. Biol.* **92**, 714.
Lange, Y., and Slayton, J. M., 1982, *J. Lipid Res.* **23**, 1121.
Lee, J. C.-M., and Discher, D. E., 2001, *Biophys. J.* **81**, 3178.
Lee, J. C-M., Wong, D. T., and Discher, D. E., 1999, *Biophys. J.* **77**, 853.
Leibler, S., and Maggs, A. C., 1990, *Proc. Natl. Acad. Sci. USA* **87**, 6433.
Lenormand, G., Hénon, S., Richert, A., Siméon, J., and Gallet, F., 2001, *Biophys. J.* **81**, 43.
Lenormand, G., Hénon, S., Richert, A., Siméon, J., and Gallet, F., 2003, *Biorheology* **40**, 247.
Li, J., Dao, M., Lim, C. T., and Suresh, S., 2005, *Biophys. J.* **88**, 3707.
Lichtman, M., Beutler, E., Kaushansky, K., Kipps, T., Seligsohn, U., and Prchal, J., 2005, *Williams Hematology*. McGraw-Hill, New York.
Lim, C. T., Dao, M., Suresh, S., Sow, C. H., and Chew, K. T., 2004, *Acta Mater.* **52**, 1837.
Lim, C. T., Dao, M., Suresh, S., Sow, C. H., and Chew, K. T., 2004, *Acta Mater.* **52**, 4065.
Lim, G. H. W., 2003, PhD thesis, Simon Fraser University, Burnaby, BC, Canada (unpublished).
Lim, G. H. W., Mukhopadhyay, R., and Wortis, M., 2002, *Proc. Natl. Acad. Sci. USA* **99**, 16766.
Lim, G. H. W., Wortis, M., and Mukhopadhyay, R., 2001, in *Proceedings of the 2001 Bioengineering Conference*, edited by R. D. Kamm, G. W. Schmid-Schonbein, G. A. Ateshian, and M. S. Hefzy, Vol. 50, p. 865 (Abstract only). American Society of Mechanical Engineers, New York.
Linderkamp, O., Friederichs, E., and Meiselman, H. J., 1993, *Pediatr. Res.* **34**, 688.
Linderkamp, O., and Meiselman, H. J., 1982, *Blood* **59**, 1121.
Linderkamp, O., Nash, G. B., Wu, P. Y. K., and Meiselman, H. J., 1986, *Blood* **67**, 1244.

Linderkamp, O., Wu, P. Y. K., and Meiselman, H. J., 1983, *Pediatr. Res.* **17**, 250.

Liu, S. C., and Palek, J., 1984, *J Biol. Chem.* **259**, 11556.

Liu, S. C., Palek, J., and Prchal, J. T., 1982, *Proc. Natl. Acad. Sci. USA* **79**, 2072.

Lomholt, M. A., and Miao, L., 2006, *J. Phys. A: Math. Gen.* **39**, 10323.

Mase, G. T., and Mase, G. E., 1999, *Continuum Mechanics for Engineers*. CRC Press, Boca Raton, FL.

McMillan, D. E., Mitchell, T. P., and Utterback, N. G., 1986, *J. Biomech.* **19**, 275.

Miao, L., 1992, PhD thesis, Simon Fraser University, Burnaby, BC, Canada (unpublished).

Miao, L., Fourcade, B., Rao, M., Wortis, M., and Zia, R. K. P., 1991, *Phys. Rev. A* **43**, 6843.

Miao, L., Siefert, U., Wortis, M., and Döbereiner, H.-G., 1994, *Phys. Rev. E* **49**, 5389.

Millman, R. S., and Parker, G. D., 1977, *Elements of Differential Geometry*, p. 188. Prentice-Hall, Englewood Cliffs, NJ.

Mills, J. P., Qie, L., Dao, M., Lim, C. T., and Suresh, S., 2004, *Mech. Chem. Biosys.* **1**, 169.

Mohandas, N., and Evans, E. A., 1994, *Annu. Rev. Biophys. Biomol. Struct.* **23**, 787.

Mouritsen, O. G., and Andersen, O. S., 1998, *In Search of a New Biomembrane Model*. Munksgaard, Copenhagen. (*Biol. Skr. Dan. Vid. Selsk.* **49**)

Mukhopadhyay, R., Lim, G. H. W., and Wortis, M., 2002, *Biophys. J.* **82**, 1756.

Nash, G. B., and Meiselman, H. J., 1983, *N. Y. Acad. Sci.* **416**, 255.

Newman, M. E. J., and Barkema, G. T., 1999, *Metropolis Monte Carlo Method*. Clarendon Press, Oxford.

Ogden, R. W., 1984, *Nonlinear Elastic Deformations*. Dover, Mineola, NY.

Ou-Yang, Z.-C., and Helfrich, W., 1987, *Phys. Rev. Lett.* **59**, 2486.

Ou-Yang, Z.-C., and Helfrich, W., 1987, *Phys. Rev. Lett.* **60**, 1209.

Ou-Yang, Z.-C., and Helfrich, W., 1989, *Phys. Rev. A* **39**, 5280.

Pai, B. K., and Weymann, H. D., 1980, *J. Biomech.* **13**, 105.

Parker, K. H., and Winlove, C. P., 1999, *Biophys. J.* **77**, 3096.

Peterson, M. A., 1992, *Phys. Rev. A* **45**, 4116.

Peterson, M. A., Strey, H., and Sackmann, E., 1992, *J. Phys. II* **2**, 1273.

Ponder, E., 1948, *Hemolysis and Related Phenomena*. Grune & Stratton, New York.

Powers, T. R., Huber, G., and Goldstein, R. E., 2002, *Phys. Rev. E* **65**, 041901.

Press, W. T., Teukolsky, S. A., Vetterling, W. T., and Flannery, B. P., 1994, *Numerical Recipes in C: the Art of Scientific Computing*, 2nd ed, p. 279. Cambridge University Press, Cambridge.

Rand, R. P., 1967, *Proc. Fedn. Am. Socs. Exp. Biol.* **26**, 1780.

Rand, R. P. and Burton, A. C., 1963, *J. Cell Comp. Physiol.* **61**, 245.

Raphael, R. M., and Waugh, R. E., 1966, *Biophys. J.* **71**, 1374.

Raphael, R. M., Waugh, R. E., Svetina, S., and Žekš, B., 2001, *Phys. Rev. E* **64**, 051913-1.

Rasia, M., and Bollini, A., 1998, *Biochim. Biophys. Acta* **1372**, 198.

Reinhart, W. H., and Chien, S., 1986, *Blood* **67**, 1110.

Rodgers, W., and Glaser, M., 1991, *Proc. Natl. Acad. Sci. USA* **88**, 1364.

Rodgers, W., and Glaser, M., 1993, *Biochemistry* **32**, 12591.

Rodgers, W., and Glaser, M., 1993, in *Fluorescence Microscopic Imaging of Membrane Domains*, edited by B. Herman and J. J. Lemasters, pp. 263–283. Academic Press, San Diego.

Ruef, P., and Linderkamp, O., 1999, *Pediatr. Res.* **45**, 114.

Scheffer, L., Bitler, A., Ben-Jacob, E., and Korenstein, R., 2001, *Eur. Biophys. J.* **30**, 83.

Schmidt, C. F., Svoboda, K., Lei, N., Petsche, I. B., Berman, L. E., Safinya, C. R., and Grest, G. S., 1993, *Science* **259**, 952.

Schreier, S., Malheiros, S. V. P., 2000, *Biochim. Biophys. Acta* **1508**, 210.

Seifert, U., 1997, *Adv. Phys.* **46**, 13.

Seifert, U., Berndl, K., and Lipowsky, R., 1991, *Phys. Rev. A* **44**, 1182.

Seigneuret, M., and Devaux, P. F., 1984, *Proc. Natl. Acad. Sci. USA* **81**, 3751.

Sheetz, M. P., and Singer, S. J., 1974, *Proc. Natl. Acad. Sci. USA* **71**, 4457.

Simons, K., and Ikonen, E., 1997, *Nature* **387**, 569.

Skalak, R., Tozeren, A., Zarda, R. P., and Chen, S., 1973, *Biophys. J.* **13**, 245.

Sleep, J., Wilson, D., Simmons, R., and Gratzer, W., 1999, *Biophys. J.* **77**, 3085.

Smith, J. E., Mohandas, N., Clark, M. R., Greenquist, A. C., and Shohet, S. B., 1980, *Am. J. Hematol.* **8**, 1.

Smith, J. E., Mohandas, N., and Shohet, S. B., 1982, *Am. J. Vet. Res.* **43**, 1041.

Spivak, M., 1965, *Calculus on Manifolds*. W. A. Benjamin, New York.

Spivak, M., 1979, *A Comprehensive Introduction to Differential Geometry*. Publish or Perish Press, Berkeley, CA.

Stadler and Linderkamp, 1989, *Microvasc. Res.* **37**, 267.

Steck, T. L., 1989, in *Cell Shape: Determinants, Regulation, and Regulatory Role*, edited by W. D. Stein and F. Bronner, pp. 205–246. Academic Press, San Diego.

Steck, T. L., Ye, J., and Lange, Y., 2002, *Biophys. J.* **83**, 2118.

Stokke, B. T., Mikkelsen, A., and Elgsaeter, A., 1986, *Eur. Biophys. J.* **13**, 203.

Stokke, B. T., Mikkelsen, A., and Elgsaeter, A., 1986, *Eur. Biophys. J.* **13**, 219.

Strange, K. (ed.), 1994, *Cellular and Molecular Physiology of Cell Volume Regulation*. CRC Press, Boca Raton, FL.
Strey, H., Peterson, M., and Sackmann, E., 1995, *Biophys. J.* **69**, 478.
Svetina, S., Brumen, M., and Žekš, B., 1985, *Stud. Biophys.* **110**, 177.
Svetina, S., Ottova-Leitmanova, A., and Glaser, R., 1982, *J. Theor. Biol.* **94**, 13.
Svetina, S., and Žekš, B., 1983, *Biomed. Biochim. Acta* **42**, S86.
Svetina, S., and Žekš, B., 1985, *Biomed. Biochim. Acta* **44**, 979.
Svetina, S., and Žekš, B., 1989, *Eur. Biophys. J.* **17**, 101.
Svetina, S., Žekš, B., Waugh, R. E., and Raphael, R. M., 1998, *Eur. Biophys. J.* **27**, 197.
Svoboda, K., Schmidt, C. F., Branton, D., and Block, S. M., 1992, *Biophys. J.* **63**, 784.
Tolédano, J.-C., and Tolédano, P., 1987, *The Landau Theory of Phase Transitions*. World Scientific Publishing Co., Singapore.
Tuvia, S., Levin, S., Bitler, A., and Korenstein, R., 1998, *J. Cell Biol.* **141**, 1551.
Van Dort, H. H., Moriyama, R., and Low, P. S., 1998, *J. Biol. Chem.* **273**, 14819.
Van Hemmen, J. L., and Leibold, C., 2007, *Phys. Repts.* **444**, 51.
Vertessy, B. G., and Steck, T. L., 1989, *Biophys. J.* **55**, 255.
Waugh, R. E., 1992, *Clin. Hemorheol.* **12**, 649.
Waugh, R. E., 1996, *Biophys. J.* **70**, 1027.
Waugh, R. E., and Bauserman, R. G., 1995, *Ann. Biomed. Eng.* **23**, 308.
Waugh, R. E., and Evans, E. A., 1976, *Microvasc. Res.* **12**, 291.
Waugh, R. E., Mohandas, N., Jackson, C. W., Mueller, T. J., Suzuki, T., and Dale, G. L., 1992, *Blood* **79**, 1351.
Welti, R., and Glaser, M., 1994, *Chem. Phys. Lipids* **73**, 121.
Wong, P., 1999, *J. Theor. Biol.* **196**, 343, and references therein.
Workman, R. F., and Low, P. S., 1998, *J. Biol. Chem.* **273**, 6171.
Wortis, M., 1998, *Biol. Skr. Dan. Vidensk. Selsk.* **49**,59.
Wortis, M., 2001, in *56th ACS Northwest Regional Meeting*, p. 93 (Abstract only). American Chemical Society, Seattle.
Wortis, M., and Evans, E. A., 1997, *Phys. Canada* **53**, 281.
Yeung, A., and Evans, E., 1995, *J. Phys. II France* **5**, 1501.
Zarda, P. R., 1974, PhD thesis, Columbia University, New York (unpublished).
Zarda, P. R., Chien, S., and Skalak, R., 1977, *J. Biomech.* **10**, 1977.
Zeman, K., Engelhard, H., and Sackmann, E., 1990, *Eur. Biophys. J.* **18**, 203.
Zilker, A., Engelhardt, H., and Sackmann, E., 1987, *J. Phys. (France)* **48**, 2139.
Zilker, A., Ziegler, M., and Sackmann, E., 1992, *Phys. Rev. A* **46**, 7998.

Index

a

acanthocyte 107
ADE model 110
area difference ΔA 91, 112
area modulus K_A 91, 109
area-difference energy F_{ad} 110, 112
area-difference modulus $\bar{\kappa}$ 91, 112, 204
area-difference parameter $\bar{m}_0$ 91, 154, 182, 198
atomic force microscopy 28
atomistic simulation 6, 61
– all-atom 23
– assumptions 6
– forcefield 7
– radial distribution factor (rdf) 35, 37
– united-atom 23
axisymmetric discocyte (AD) 150, 168, 169, 189, 190
axisymmetric shape equations 135, 239
axisymmetric stomatocyte (AS) 150, 165, 167

b

bending energy F_b 214
bending energy of the plasma membrane F_b 111

bending modulus κ_b 91, 94, 110, 204, 207, 215, 231
bilayer thickness D 91
bilayer–couple hypothesis (Sheetz-Singer) 98, 193
Boltzmann weighted sampling 50

c

Cahn–Hilliard equation 62
cholesterol 4, 38, 62, 65
Christoffel symbols 138, 220, 223, 240
coarse grain model 59–62
Codazzi relations 221, 224
codocyte 107
computational area modulus K_A* 91, 141
computational osmotic modulus K_V* 91, 141
constraint energies 108, 141
continuous transitions 96, 152, 153
contravariance, covariance 221
covariant derivatives 223, 224
curvature tensor **C** 110, 221, 225
cytoskeleton 90

d

deformation matrix **M** 216, 236, 238
detailed balance 50
differential geometry 218
dimensionless variables 118
discocyte 85
discontinuous transitions 96, 152, 153
discretization 139

e

echinocyte (E) 85, 105, 150, 168, 171–175, 189, 190
effective area difference 113
effective spontaneous curvature $\overline{C}_0$ 91, 94, 113
elastic energy density f_{ms} 115, 216
elastic length scale 91, 94, 119
elastic stress tensor 137, 139
elliptocyte 107
energy of the membrane skeleton 113, 145
energy of the plasma membrane 110, 112, 143
erythrocyte 85
– structure 89

f

flip-flop 103, 197
fluid mosaic model 4

g

Gauss's law 131, 137, 225
Gauss–Bonnet theorem 111, 136, 234, 235
Gaussian bending energy F_g 110, 234
Gaussian curvature K 110, 221
Gaussian modulus κ_g 91, 110, 204, 215, 235

Gouy–Chapman theory 31

h

2 H–NMR 27
heat capacity 59
Helfrich equation 136, 240
Helfrich model 110, 120, 132
Helfrich stress tensor 133, 229, 232
Helfrich torque 132, 231
history 88, 99, 120
hyperelastic material 114
hysteresis 93, 106, 151, 152, 185

i

intrinsic torque 131, 234

k

keratocyte 107
knizo-echinocyte (KE) 150, 168, 173, 175–179
knizocyte (K) 87, 107, 150, 168, 173, 176

l

Laplace-Beltrami operator 224
lipid asymmetry 103, 197
lipid bilayer 2, 4
– ζ–potential 31
– area per lipid (A_l) 13, 20
– atomistic simulation 5
– coarse grain simulation 60
– configurational bias Monte Carlo 52
– dipole potential 33
– domain 59, 62, 75
– electron density 26
– electrophoretic experiment 31
– heterogeneous 36
– hydrocarbon volume (V_c) 13
– lipid diffusion 21
– lipid volume (V_l) 13

– molecular dynamics   9
– – forcefield   9–12
– – water model   23
– order parameter   28, 62, 66
– partial molecular area   38
– partial molecular volume   38
– phase separation   59, 73
– phase transition   5, 67
– specific heat   71
– surface potential   33
lysis   92, 106

m

Maier Saupe model   62
Marčelja model   64
Markov chain   50
mean curvature H 110, 221
mean field approximation   63, 64
mechanics of the membrane skeleton   137
mechanics of the plasma membrane   125
membrane energies   107
membrane skeleton   90, 104, 238
– strain distribution   187–189
membrane-energy minimization   93
membrane-shape mechanics   123
membrane-skeleton thickness   91
membranes with internal stresses   129
metastability   93
metric tensor g 129, 219
Metropolis algorithm   48
molecular dynamics   8
Monte Carlo energy minimization   146

n

neutral surface   112, 230
non-axisymmetric discocyte (NAD)   87, 106, 167, 168, 170
non-axisymmetric stomatocyte (NAS)   150, 163–166
nonlinear elastic coefficients   91, 116, 195, 210
nonlocal bending modulus   112, 204
Nose–Hoover thermostat   18

o

osmotic modulus K_V 91, 109
Ou-Yang equations   135, 233, 240

p

parallel tempering   56
phase diagrams   95, 99, 194
phase space   95
phase-trajectory diagrams   154–159, 161
phase-trajectory´diagrams   160
phospholipids   3
– DMPC   62
– DOPC   13, 20, 29, 38
– DPPC   3, 13, 20, 38, 65
– PI   47
– POPC   13, 20, 38
plasma membrane   89, 102
principal curvatures C_i 110, 111, 127, 221, 225
principal extension ratios λ_i 114, 138, 217, 237
principal stresses τ_i 138

r

red-cell area   90, 91, 204, 205
red-cell volume   90, 91, 204, 205
reduced volume v 90, 91

reference shape S_0 94, 114, 116, 117, 183
relaxed area difference ΔA_0 91, 111
reparametrization invariance   118
replica exchange   56
reticulocyte   89, 107, 184

s

scaling   118
SCMFT algorithm   66
SDE sequence of transformations   86, 87, 98, 102, 105, 182, 185, 196
semi-Gibbs ensemble Monte Carlo   57
shape classes   86, 95, 149, 162, 182
shape-change agents   86, 193, 196
shape-energy functional $F[S]$ 94, 107
shear modulus μ 91, 94, 115, 195, 206, 208
spectrin   104
sphering   92, 121
sphero-echinocyte   105
sphero-stomatocyte   105
sphingolipids   3
– ceramide   20
– sphingomylin   20, 29
spiculated shapes (SS)   150, 177, 179–181, 190
spontaneous curvature C_0 91, 110, 215

spontaneous-curvature model   110, 120
stability boundaries   95, 152
stability diagrams   99, 162, 183
stomatocyte   85, 105
strain invariants α and β 115, 218, 237
stress tensor   130, 217, 228, 237, 238
stretch modulus K_α 91, 94, 115, 195, 206, 208
structure of the cell membrane   102, 103

t

tangent vectors $\mathbf{Y}$ 219
thermal fluctuations   93, 153, 154, 185, 189
time dependent Ginzburg–Landau equation   65
torque   129, 131, 234
trajectory   151, 153, 194
triangular stomatocyte   87, 107, 164

u

universality   86, 98

v

van't Hoff relation   91, 109, 210

w

Weingarten relation   221, 224

x

X-ray form factor   13, 26